Inflammation and Oral Cancer

Inflammation and Oral Cancer

From Bench to Bedside

Edited by

Hiroyuki Tomita
Associate Professor, Department of Tumor Pathology,
Gifu University Graduate School of Medicine,
Gifu, Japan

ACADEMIC PRESS
An imprint of Elsevier

Academic Press is an imprint of Elsevier
125 London Wall, London EC2Y 5AS, United Kingdom
525 B Street, Suite 1650, San Diego, CA 92101, United States
50 Hampshire Street, 5th Floor, Cambridge, MA 02139, United States
The Boulevard, Langford Lane, Kidlington, Oxford OX5 1GB, United Kingdom

Copyright © 2022 Elsevier Inc. All rights reserved.

No part of this publication may be reproduced or transmitted in any form or by any means, electronic or mechanical, including photocopying, recording, or any information storage and retrieval system, without permission in writing from the publisher. Details on how to seek permission, further information about the Publisher's permissions policies and our arrangements with organizations such as the Copyright Clearance Center and the Copyright Licensing Agency, can be found at our website: www.elsevier.com/permissions.

This book and the individual contributions contained in it are protected under copyright by the Publisher (other than as may be noted herein).

Notices
Knowledge and best practice in this field are constantly changing. As new research and experience broaden our understanding, changes in research methods, professional practices, or medical treatment may become necessary.

Practitioners and researchers must always rely on their own experience and knowledge in evaluating and using any information, methods, compounds, or experiments described herein. In using such information or methods they should be mindful of their own safety and the safety of others, including parties for whom they have a professional responsibility.

To the fullest extent of the law, neither the Publisher nor the authors, contributors, or editors, assume any liability for any injury and/or damage to persons or property as a matter of products liability, negligence or otherwise, or from any use or operation of any methods, products, instructions, or ideas contained in the material herein.

Library of Congress Cataloging-in-Publication Data
A catalog record for this book is available from the Library of Congress

British Library Cataloguing-in-Publication Data
A catalogue record for this book is available from the British Library

ISBN 978-0-323-88526-3

For information on all Academic Press publications
visit our website at https://www.elsevier.com/books-and-journals

Publisher: Stacy Masucci
Acquisitions Editor: Rafael E. Teixeira
Editorial Project Manager: Pat Gonzalez
Production Project Manager: Kiruthika Govindaraju
Cover Designer: Miles Hitchen

Typeset by STRAIVE, India

Contents

Contributors	xi
Preface	xiii

1. Epidemiology of oral cancer and its relationship with inflammation

Katsuaki Bunai

1	**Introduction**	1
2	**Oral anatomy and favorite site**	2
3	**Histological type**	2
4	**Age and sex**	3
5	**Incidence and mortality**	4
6	**Risk factors**	6
	6.1 Tobacco and oral cancer	7
	6.2 Alcohol and oral cancer	7
	6.3 Betel quid and oral cancer	7
	6.4 Virus HPV, EBV	7
	6.5 Oral microbiome	8
	6.6 Candida	9
7	**Genetic mutation and signal transduction**	9
	7.1 EGFR pathway	11
	7.2 RAS and MAPK pathway	11
	7.3 PI3K, AKT, and mTOR pathway	11
	7.4 NF-κB pathway	12
	7.5 Jak and STAT pathway	12
	7.6 WNT pathway	12
	7.7 NOTCH pathway	12
	7.8 Hippo pathway	13
	7.9 PD-1/PD-L1 pathway	13
8	**Inflammation and cancer**	14
	8.1 Cancer microenvironment	14
	8.2 Cancer microenvironment and immune checkpoint mechanism	15
	References	15

2. Diagnosis of oral squamous cell carcinomas and precancerous lesions

Keisuke Nakano and Hitoshi Nagatsuka

1	**Introduction**	19
2	**OPMDs**	20
	2.1 Definition and classification of OPMDs	20
	2.2 Leukoplakia	20
	2.3 Erythroplakia	21
	2.4 Oral lichen planus	21
	2.5 Oral submucous fibrosis	22
	2.6 Oral candidiasis	22
	2.7 Oral syphilis	24
	2.8 Discoid lupus erythematosus	24
	2.9 Proliferative verrucous leukoplakia	25
3	**Oral epithelial dysplasia**	25
	3.1 Definition and classification of OED	25
	3.2 Immunohistochemistry (IHC) of OED	26
4	**Oral cancer**	29
	4.1 Epidemiology of oral cancer	29
	4.2 Clinical findings of oral cancer	29
	4.3 Oral cancer histology	30
	4.4 Squamous cell carcinoma	30
	4.5 Grade classification of SCC	31
	4.6 IHC for SCC	32
	4.7 CIS	32
5	**Subtype of oral SCC**	35
	5.1 Basaloid SCC	35
	5.2 Spindle cell SCC	35
	5.3 Adenosquamous carcinoma	36
	5.4 Carcinoma cuniculatum	36
	5.5 Verrucous SCC	37
	5.6 Lymphoepithelial carcinoma	38
	5.7 Papillary SCC	38
	5.8 Acantholytic SCC	39
6	**Conclusion**	39
	Conflict of interest statement	39
	References	40

3. Oral cytology in Japan: Its useful approach and criteria for early detection of carcinoma and precursor lesions

Motohiko Nagayama, Akinori Ihara, and Yoichi Tanaka

1	**Introduction**	43
2	**Procedure of cell collection**	44
3	**The guideline diagnostic criteria for oral mucosal exfoliative cytology (OMEC)**	45
4	**Appropriate evaluation of specimens**	45

Contents **vii**

5 Cytological classification 45
  5.1 Correspondence between the histological grading system for epithelial dysplasia and the cytodiagnostic criteria 49
  5.2 Approach of oral mucosal exfoliative cytology and diagnostic criteria with flow chart 51
  5.3 Differential cytodiagnosis 52
  5.4 Usefulness, limitations, and future prospects of the LBC method 52
  References 53

4. Imaging findings of oral cancers

Hiroki Kato and Masayuki Matsuo

1 Introduction 55
2 Anatomy of the oral cavity 56
3 Staging of oral cancer 56
4 Imaging modalities and protocols in oral cancer 58
  4.1 CT 58
  4.2 MRI 59
  4.3 ^{18}F-fluorodeoxyglucose positron emission tomography (PET)/CT 60
5 Imaging findings of oral cancer at each subsite 60
  5.1 Lip 60
  5.2 Oral tongue 61
  5.3 Buccal mucosa and gingiva 64
  5.4 Floor of the mouth 65
  5.5 Hard palate 67
  5.6 Retromolar trigone 69
6 Patterns of spread of oral cancer 71
  6.1 Bone invasion 71
  6.2 Perineural spread 71
  6.3 Lymph node metastasis 73
  6.4 Distant metastasis 76
7 Conclusion 76
  References 76

5. What is epigenetics?

Keizo Kato

1 Foreword 79
2 Methylation modification of DNA 80
3 Writing with DNA methyltransferases 82
4 Reversibility of methylation 82
5 DNA methylation in cancer 83
6 CpG island standards 83
7 Activation of oncogenes 83
8 Silencing of tumor suppressor genes 84
9 Causes of methylation abnormalities 84

viii Contents

9.1	Exogenous factors	84
9.2	Endogenous factors	84
10	**Methylation detection methods**	85
10.1	Analysis of individual gene regions	85
10.2	Comprehensive genome analysis	87
11	**Methylation of various carcinomas**	87
11.1	Oral cancer	87
11.2	Nonoral cancers	89
12	**Possibility of application in cancer treatment**	90
13	**Hypermethylation detection in blood and body fluids and assay methods for clinical application**	91
14	**Toward the realization of methylation therapies**	92
15	**Oral cancer treatment**	93
16	**Future perspectives**	98
	References	98

6. Role of autophagy in dysregulation of oral mucosal homeostasis

Madoka Yasunaga, Masahiro Yamaguchi, Kei Seno, Mizuki Yoshida, and Jun Ohno

1	**Introduction**	101
2	**Oral mucosal homeostasis**	102
2.1	Oral mucosal architecture	102
2.2	Maintenance of oral mucosal homeostasis	102
3	**Autophagy**	102
3.1	Five steps in autophagosome formation	102
3.2	Signaling pathways of autophagy	104
3.3	Effect of autophagy on cellular differentiation	104
3.4	Regulation of osteogenesis by autophagy	105
4	**Effect of autophagy on oral infection**	106
4.1	Oral infection	106
4.2	What is xenophagy?	107
4.3	Xenophagy in periodontal infection	108
5	**Effect of autophagy on aging and cellular senescence**	109
5.1	Aging and autophagy	109
5.2	Cellular senescence and autophagy	110
6	**Autophagy in cancer**	113
6.1	Effects of autophagy on tumor metabolism	113
6.2	Cancer stem cells (CSCs)	114
6.3	Tumor-suppressing function of autophagy	115
6.4	Tumor-promoting function of autophagy	116
6.5	Role of autophagy in tumor cell dissemination and metastasis	116
6.6	Autophagy as a therapeutic tool for cancer	117
6.7	Therapeutic targets at the early stages of autophagy	117
6.8	Therapeutic targets at the late stages of autophagy	117

Contents **ix**

7 Conclusion 118
Acknowledgments 118
Conflict of interest 118
References 118

7. Oral mucosal graft-versus-host disease and its possibility of antitumor effects

Kei Seno, Madoka Yasunaga, Nana Mori-Yamamoto, and Jun Ohno

1 Introduction 127
2 **What is GVHD?** 127
    2.1 aGVHD 128
    2.2 cGVHD 129
3 **Oral GVHD** 129
    3.1 Oral aGVHD 130
    3.2 Oral cGVHD 131
4 **Immunopathophysiology of oral GVHD** 133
    4.1 Contribution of animal models for pathophysiology of GVHD 133
    4.2 A three step model of aGVHD pathophysiology 134
    4.3 Pathophysiology of mucocutaneous aGVHD 136
5 **Graft-versus-tumor as antitumor effect of GVHD in oral squamous cell carcinoma** 142
6 **Conclusion** 144
Acknowledgments 144
Conflict of interest 144
References 144

8. Standardization of sentinel lymph node biopsy in oral cancer

Kazuhisa Ishida

1 Introduction 151
2 Identification of sentinel lymph nodes 152
3 Methods for the diagnosis of sentinel lymph node metastasis 154
4 Evidence-based medicine applied to sentinel lymph node biopsy (SLNB) 158
5 Summary 159
References 160

9. Overview of radiotherapy for oral cavity cancer

Chiyoko Makita, Masaya Ito, Hirota Takano,
Tomoyasu Kumano, and Masayuki Matsuo

1 Introduction 165
2 General principles 166
    2.1 Definitive radiotherapy for early-stage oral cavity cancer 166
    2.2 Postoperative radiotherapy and chemoradiotherapy 166

x Contents

2.3	Primary radiotherapy or chemoradiotherapy without surgery	169
2.4	Palliative radiotherapy	169
2.5	Management for the toxicities of radiotherapy	170
3	**Radiation technique**	170
3.1	EBRT	170
3.2	Brachytherapy	171
3.3	IMRT	172
3.4	IGRT	173
3.5	Particle beam therapy	175
4	**Conclusion**	176
	Acknowledgments	177
	Conflict of interest statement	177
	References	177

10. Management of cancer treatment-induced oral mucositis

Akio Suzuki

1	**Introduction**	183
2	**Frequency of oral mucositis**	184
3	**Pathogenesis of anticancer agent and radiotherapy-induced oral mucositis**	184
4	**Management of oral mucositis**	185
4.1	Oral care	185
4.2	Cryotherapy	186
4.3	Pharmacotherapy for protection or treatment of chemotherapy- and/or radiotherapy-induced oral mucositis	186
5	**Effect of polaprezinc on chemotherapy- and/or radiotherapy-induced oral mucositis**	191
	References	192

11. Perspectives in research on oral squamous cell carcinoma

Hiroyuki Tomita

Index   205

Contributors

Numbers in parenthesis indicate the pages on which the authors' contributions begin.

Katsuaki Bunai (1), Department of Oral and Maxillofacial Surgery, Gifu University Graduate School of Medicine, Gifu, Japan

Akinori Ihara (43), Ihara-Machinaka Dental Clinic, Numazu, Shizuoka, Japan

Kazuhisa Ishida (151), Oral and Maxillofacial Surgery, Graduate School of Medicine, Gifu University, Gifu, Japan

Masaya Ito (165), Department of Radiology, Gifu University Graduate School of Medicine, Gifu, Japan

Hiroki Kato (55), Department of Radiology, Gifu University Graduate School of Medicine, Gifu, Japan

Keizo Kato (79), Department of Oral and Maxillofacial Surgery, Gifu University Graduate School of Medicine, Gifu, Japan

Tomoyasu Kumano (165), Department of Radiology, Gifu University Graduate School of Medicine, Gifu, Japan

Chiyoko Makita (165), Department of Radiology, Gifu University Graduate School of Medicine, Gifu, Japan

Masayuki Matsuo (55, 165), Department of Radiology, Gifu University Graduate School of Medicine, Gifu, Japan

Nana Mori-Yamamoto (127), Section of Periodontology, Department of Odontology, Fukuoka Dental College, Fukuoka, Japan

Hitoshi Nagatsuka (19), Department of Oral Pathology and Medicine, Graduate School of Medicine, Dentistry and Pharmaceutical Sciences, Okayama University, Okayama, Japan

Motohiko Nagayama (43), Department of Oral Pathology, Asahi University School of Dentistry, Mizuho, Gifu, Japan

Keisuke Nakano (19), Department of Oral Pathology and Medicine, Graduate School of Medicine, Dentistry and Pharmaceutical Sciences, Okayama University, Okayama, Japan

Jun Ohno (101, 127), Oral Medicine Research Center, Fukuoka Dental College, Fukuoka, Japan

Kei Seno (101, 127), Section of General Dentistry, Department of General Dentistry, Fukuoka Dental College, Fukuoka, Japan

Akio Suzuki (183), Department of Pharmacy, Gifu University Hospital, Gifu, Japan

xii Contributors

Hirota Takano (165), Department of Radiotherapy, Gifu University Graduate School of Medicine, Gifu, Japan

Yoichi Tanaka (43), Department of Surgical Pathology, Tokyo Dental College Ichikawa General Hospital, Ichikawa, Chiba, Japan

Hiroyuki Tomita (199), Department of Tumor Pathology, Gifu University Graduate School of Medicine, Gifu, Japan

Masahiro Yamaguchi (101), Section of Gerodontology, Department of General Dentistry, Fukuoka Dental College, Fukuoka, Japan

Madoka Yasunaga (101, 127), Section of Orthodontics, Department of Oral Growth and Development, Fukuoka Dental College, Fukuoka, Japan

Mizuki Yoshida (101), Section of Gerodontology, Department of General Dentistry, Fukuoka Dental College, Fukuoka, Japan

Preface

In recent years, cancer research has seen rapid innovation in experimental techniques, including next-generation sequencing. In addition, the quantity of new findings, such as the influence of oral bacteria on systemic diseases, is rapidly increasing. In addition to basic medical research, clinical medicine is also undergoing rapid changes in diagnosis and treatment. Therefore, this book summarizes the wide range of basic and clinical stomatitis and oral cancer and is intended to be read by all people involved in caring for and treating the oral cavity, from general practitioners to specialist clinical doctors and other medical professionals.

The book covers a wide range of topics, focusing on oral cancer, including epidemiology, imaging, pathology, treatment, and basic research.

It is rare for a book on oral diseases to include contributions from such a diverse group of authors. In particular, this time, we have a pharmacist writing about mouth ulcers associated with anticancer drugs. This chapter is also essential for medical professionals dealing with other cancers.

Finally, I would be honored if everyone involved with caring for and treating the oral cavity, from general practitioners to specialist clinical doctors and other medical professionals, would pick up the book and read it.

Hiroyuki Tomita
Department of Tumor Pathology, Gifu University Graduate
School of Medicine,
Gifu, Japan

Chapter 1

Epidemiology of oral cancer and its relationship with inflammation

Katsuaki Bunai
Department of Oral and Maxillofacial Surgery, Gifu University Graduate School of Medicine, Gifu, Japan

1. Introduction

When searching the literature on oral cancer, we find an article entitled "Cancer of the Mouth" in *the New England Journal of Medicine*, published in 1919 [1]. Interestingly, tobacco seems to have been recognized as a causative factor for oral cancer since this period. Furthermore, although the word "inflammation" is not used, the possibility of "chronic irritation" contributing to mouth malignancy has been indicated. We can go back to an 1863 report by Rudolf Virchow with regards to the link between cancer and inflammation [2]. In general, "inflammation" and "cancer" are classified as separate diseases. However, since there are many similarities between them, there has been much debate about these distinct divisions in various fields, such as clinical medicine, epidemiology, and pathology. Approximately 150 years later, there have been many reports that support the link between inflammation and cancer. Specific, well-known relationships between inflammation and cancer include infectious diseases, such as *Helicobacter pylori* infection and gastric cancer, as well as hepatitis and liver cancer. In addition to these infectious diseases, cancers related to inflammation, such as mesothelioma caused by asbestos, and ulcerative colitis and colorectal cancer, are said to account for 15%–20% of all cancers [3].

Tobacco and alcohol are the most frequent causal factors in oral cancer, but the possibility of various factors being involved in carcinogenesis is becoming clear. Additionally, progress in research methods, especially those of molecular biology, has resulted in the discovery of molecules involved in inflammation and carcinogenesis. It has become clear that these are intricately involved and act to promote carcinogenesis, or sometimes, suppress it. In recent years,

---

Inflammation and Oral Cancer. https://doi.org/10.1016/B978-0-323-88526-3.00001-4
Copyright © 2022 Elsevier Inc. All rights reserved.

2 Inflammation and oral cancer

the advent of microarrays and next-generation sequencers has enabled comprehensive gene searches and batch searches by bacterial flora. Immune checkpoint molecules, such as PD-1/PD-L1, have also been discovered, and the system that controls inflammation and immunity is becoming clear. However, in contrast to the results of these studies, the number of patients with oral cancer continues to grow worldwide. GLABOCAN predicts that the number of patients will increase by a factor of 1.46 in the next 20 years [4]. A characteristic of oral cancer is that the number of patients varies considerably from region to region, and these characteristics will be described in this chapter.

2. Oral anatomy and favorite site

Oral anatomy has various definitions, depending on the field. The well-known TNM (tumor, nodes, metastasized) classification of the Union for International Cancer Control (UICC) is applied only to carcinoma and is widely used in defining and classifying oral cancer. The eighth edition, published in 2017, roughly divided the oral cavity into six parts: buccal mucosa, upper alveolar and gingiva, inferior alveolar and gingiva, hard palate, tongue, and floor of the oral cavity. Among these, the buccal mucosa is subdivided into the mucosa of the upper and lower lips, mucosa of the cheek, posterior acetabulum, and upper/lower buccal gingival sulcus. The tongue is subdivided into the dorsum of the tongue after the circumvallate papilla, tongue margin, and undersurface of the tongue. In this classification, the lips are separated from the oral cavity, although in Japan, cancer of the lips is often treated as oral cancer (Table 1). According to a report by the Japan Society for Head and Neck Cancer Center Registry Committee in 2017, tongue cancer ranked first among oral cancers, with 53.5% of the total cases, when classified by site. This is followed by the inferior alveolar and gingiva (14.6%), the upper alveolar and gingiva (9.5%), floor of the mouth (9.2%), buccal mucosa (9.1%), and the hard palate (3.1%). These results were generally consistent between 2011 and 2017 [5].

3. Histological type

The oral mucosa consists of stratified squamous epithelium. Therefore, there are many squamous cell carcinomas, accounting for more than 90% of oral cancers. These are common throughout the world, and the histological types are often well-differentiated to moderately differentiated, which is different from the poorly differentiated and undifferentiated types that are often found in the nasal cavity, nasopharynx, oropharynx, and so on. In the fourth edition of the WHO Classification of Head and Neck Tumors, published in 2017, subtypes of squamous cell carcinoma included basaloid squamous cell carcinoma, spindle-cell squamous cell carcinoma, adenosquamous carcinoma, carcinoma cuniculatum, verrucous squamous cell carcinoma, lymphoepithelial carcinoma,

Epidemiology of oral cancer and its relationship with inflammation **Chapter | 1**   **3**

TABLE 1 Anatomical sites and subsites and ICD-10 classification of lip and oral cancer.

Anatomical sites and subsites	ICD-10
Lip	(C00)
1. External upper lip (vermilion border)	(C00.0)
2. External lower lip (vermilion border)	(C00.1)
3. Commissures	(C00.6)
Oral cavity	(C02–06)
1. Buccal mucosa	
a. Mucosa of upper and lower lips	(C00.3, 4)
b. Cheek mucosa	(C06.0)
c. Retromolar areas	(C06.2)
d. Buccoalveolar sulci, upper and lower (vestibule of mouth)	(C06.1)
2. Upper alveolus and gingiva (upper gum)	(C03.0)
3. Lower alveolus and gingiva (lower gum)	(C03.1)
4. Hard palate	(C05.0)
5. Tongue	
a. Dorsal surface and lateral border anterior to vallate papillae (anterior two thirds)	(C02.0, 1)
b. Inferior (ventral) surface	(C02.2)
6. Floor of mouth	(C04)

papillary squamous cell carcinoma, and acantholytic squamous cell carcinoma. Other cancers included adenocarcinoma, represented by salivary gland cancer, and malignant melanoma. According to Japanese statistics, a report by the Japan Society for Head and Neck Cancer Center Registry Committee indicated that 94% were squamous cell carcinomas (95% if verrucous carcinomas are included), followed by 1% of mucoepidermoid carcinomas and adenoid cystic carcinomas (Table 2). These results were generally consistent between 2011 and 2017 [5–11].

4. Age and sex

The male to female ratio is approximately 2:1, and this is similar across many countries. The predominant age was 50 years or older, especially 60–80 years. Although some countries had different proportions, it is said that luxury items have a large impact, and there may be a high proportion of men in areas where smoking tobacco, especially betel quid, is a common habit. The gender gap also tends to decrease in developed countries like Japan and the United States. These differences are thought to be the result of a growing culture of smoking cessation due to the increasing health consciousness of society.

4 Inflammation and oral cancer

TABLE 2 Histological distribution of oral cancer.

	Histological distribution					Total cases
	1	**2**	**3**	**4**	**5**	
2017	SCC (3174)	VC (37)	MC (32)	ADC (30)	MM (15)	3366
2016	SCC (2476)	VC (32)	MC (26)	ADC (24)	MM (20)	2919
2015	SCC (2135)	VC (27)	MC, ADC (25)		MM (19)	2285
2014	SCC (2314)	VC (34)	ADC (28)	MC (24)	MM (19)	2505
2013	SCC (2075)	VC (33)	ADC (19)	MC (15)	MM (7)	2202
2012	SCC (1302)	MC (15)	ADC (14)	VC (11)	MM (6)	1381
2011	SCC (995)	ADC (14)	MC (8)	VC (7)	OS (5)	1064

Squamous cell carcinoma (SCC), verrucous carcinoma (VC), mucoepidermoid carcinoma (MC), adenoid cystic carcinoma (ADC), malignant melanoma (MM), osteosarcoma (OS).
Source: Japan Society for Head and Neck Cancer Center Registry Committee [5–11].

5. Incidence and mortality

With regards to the global incidence of oral cancer, there were an estimated 19,292,789 new patients with cancer of all kinds, according to GLOBOCAN of the International Agency for Research on Cancer (IARC). Among these, there were 377,713 patients with lip and oral cancer, accounting for 2.0% of all cancer patients [4]. No major changes were seen until around 2008, although the number of new oral cancer patients has increased by about 100,000 in the last 10 years. Furthermore, the number of people with oral cancer is expected to increase by 170,000 in the next 20 years. From a global perspective, these results show that the number of patients with oral cancer has increased by a factor of 1.41 from 2000 to 2020, and will increase by a factor of 1.46 in the next 20 years [4, 12–16] (Fig. 1).

Meanwhile, the number of deaths is approximately half the number of new oral cancer patients (0.46–0.5), and this ratio has not changed over the past 20 years. Furthermore, the frequency of oral cancer varies from region to region. First, a comparison of the number of patients by region shows the following: Asia, 248,360 people (65.8%); Europe, 65,279 people (17.3%); North America, 27,469 people (7.3%); Latin America and Caribbean, 17,888 people (4.7%); Africa, 14,286 people (3.8%); and Oceania, 4431 people (1.2%) (Fig. 2). Comparisons of the incidence of cancer by site show the following: Melanesia in Oceania has the highest incidence in the world, with oral cancer accounting

Epidemiology of oral cancer and its relationship with inflammation **Chapter | 1** **5**

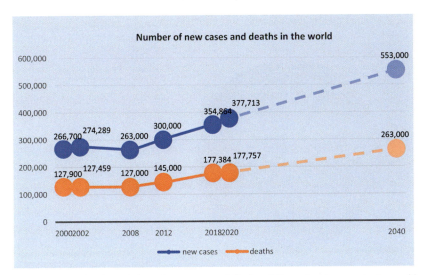

FIG. 1 Graph showing number of new cases and deaths of lip and oral cancer in the world, both sexes combined, over the period 2000–2020, and the projected future burden of cancer in 2040 based on demographic projections. *Source: Sung H, Ferlay J, Siegel RL, et al. Global cancer statistics 2020: GLOBOCAN estimates of incidence and mortality worldwide for 36 cancers in 185 countries. CA Cancer J Clin. 2021; Parkin DM, Bray F, Ferlay J, et al. Estimating the world cancer burden: Globocan 2000. Int J Cancer. 2001;94(2):153–156; Parkin DM, Bray F, Ferlay J, et al. Global cancer statistics, 2002. CA Cancer J Clin. 2005;55(2):74–108; Ferlay J, Shin HR, Bray F, et al. Estimates of worldwide burden of cancer in 2008: GLOBOCAN 2008. Int J Cancer. 2010;127(12):2893–2917; Ferlay J, Soerjomataram I, Dikshit R, et al. Cancer incidence and mortality worldwide: sources, methods and major patterns in GLOBOCAN 2012. Int J Cancer. 2015;136(5):E359–E386; Bray F, Ferlay J, Soerjomataram I, et al. Global cancer statistics 2018: GLOBOCAN estimates of incidence and mortality worldwide for 36 cancers in 185 countries. CA Cancer J Clin. 2018;68(6):394–424.*

for 8.3% of all cancers, particularly in Papua New Guinea, where lip and oral cancers account for 10.9% of all cancers in the area. Following these incidences are high occurrences in South Asia (India, Sri Lanka, Pakistan, Maldives, etc.). There also tend to be high incidences in Eastern Europe (Hungary, Slovakia, Slovenia, etc.), parts of Western Europe (Northern France, Portugal, etc.), Oceania (Australia/New Zealand), East Asia (Taiwan, etc.), and Latin America and parts of the Caribbean (Brazil, Uruguay, Puerto Rico, etc.). It is said that luxury items have a major influence on these extreme biases in the incidence rate, with betel quid, tobacco, and alcohol having the largest influences. Oral cancer ranks first in all cancer rates for men, particularly in countries where betel quid consumption is a habit, such as in India, Sri Lanka, Pakistan, and Papua New Guinea. In Japan, the incidence of oral cancer is 1%–2% of all cancers, which is the same as the proportion reported by GLOBOCAN. This accounts for 20%–30% of head and neck cancers.

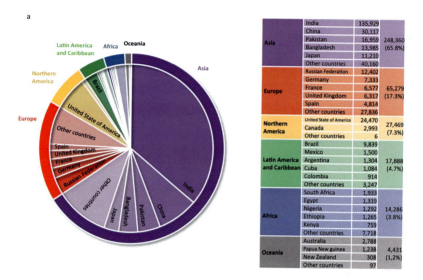

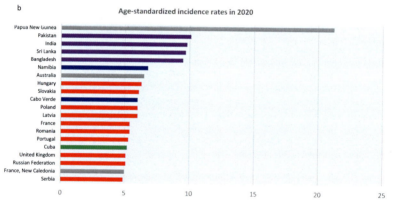

FIG. 2 Graph showing: (A) number of new cases and (B) age-standardized incidence (per 100,000 population) of lip and oral cancer. The new cases list the top five countries in each continent. *Source: GLOBOCAN 2020 (Sung H, Ferlay J, Siegel RL, et al. Global cancer statistics 2020: GLOBOCAN estimates of incidence and mortality worldwide for 36 cancers in 185 countries. CA Cancer J Clin. 2021).*

6. Risk factors

The most well-known risk factors for oral cancer are tobacco, alcohol, and betel quid. These are supported by extensive statistics in many articles. Furthermore, tobacco, alcohol, and betel quid are independent risk factors, but they also have synergistic effects and have been found to significantly increase the risk of oral cancer when used together [17]. Additionally, although rare in oral cancer, it has been suggested that viral infections such as HPV and EBV, fungal infections such as candida, and changes in the bacterial flora in the oral cavity may be clues to the development of oral cancer.

6.1 Tobacco and oral cancer

Smoking is the best-known risk factor for all cancers. It is also the most frequent causal and independent risk factor for oral cancer. The International Head and Neck Cancer Epidemiology (INHANCE) Consortium report summarized 19 research reports and performed statistical analysis on 13,935 head and neck cancer patients and 18,691 control individuals [18]. This report included 4110 patients with oral cancer, and the frequency of oral cancer increased in proportion to the frequency and duration of smoking. Meanwhile, considering the site in the head and neck, the nasopharynx and larynx showed high odds ratios, while the oral cavity and mesopharynx showed slightly lower values. Smoking rates have been declining in some areas in recent years due to increased health awareness, although there are still more than 1 billion smokers worldwide.

6.2 Alcohol and oral cancer

Alcohol is a risk factor that has deep causal relationships with oral cancer. The INHANCE Consortium report summarized 15 research reports and performed a statistical analysis on 9107 head and neck cancer patients and 14,219 control individuals [19]. This report included 572 cases of oral cancer, and the frequency of oral cancer increased in proportion to the increase in alcohol consumption. Alcohol consumption is increasing worldwide when compared to tobacco consumption, and due to its versatility, it continues to be used in the medical field, as represented by mouthwash.

6.3 Betel quid and oral cancer

The betel palm is a plant widely known across Asia. Although it is a highly regional risk factor, areas with widespread use of it have a high incidence of oral cancer. It is often chewed as a luxury item, but it also has a history of being used in Chinese medicine and as an anthelmintic [20]. Betel quid production continues to expand, despite reports of the health hazards associated with its consumption.

6.4 Virus HPV, EBV

Human papillomavirus (HPV) has been studied since it was first reported with regards to cervical cancer in 1983 [21]. Some articles suggest an association with oral cancer, although many reports suggest a strong association with oropharyngeal cancer. Additionally, HPV-positive cancer in the head and neck region is the second most common in the tongue base after the oropharyngeal region, and the interpretation may change significantly depending on the region under which HPV-positive cancer in the tongue base is classified [22]. HPV-positive cancers found in the oropharyngeal region are considered to have a good prognosis. If research results on cervical cancer are included, more than

8 Inflammation and oral cancer

100 HPV genotypes have been identified to date; among these, HPV16 and 18 have been clarified as high-risk types for head and neck cancer [23]. These high-risk HPVs contain two oncogenes (E6 and E7), and HPV-infected cells replicate viral DNA in the nucleus. The E6 protein acts on ubiquitin ligase and promotes the ubiquitination of p53 to suppress apoptosis. Additionally, the E7 protein works to promote the cell cycle by inhibiting the binding of Rb and E2F and suppressing the functions of p21 and p27.

It is thought that cancer occurs due to the accumulation of these genetic abnormalities. It is also well known that p16 expression is enhanced in a compensatory manner by inactivating p53 and Rb due to HPV infection and is used as a marker for HPV infection. The association between the Epstein–Barr virus (EBV) and cancer has been reported before HPV, and this is called lymphoepithelial carcinoma in the oral cavity [24, 25]. Most of these are found in the nasopharyngeal and salivary gland tissue, and they are quite rare in the oral cavity [26]. With regards to the areas of incidence, this is a form of cancer with large regional differences, with many reports in China, Southeast Asia, and North Africa. Additionally, EBV-positive cancer found in the oral cavity is thought to have a favorable prognosis.

6.5 Oral microbiome

Research on oral bacteria and oral cancer has been ongoing for several years. However, there has been no common recognition of its direct involvement in carcinogenesis. It may have been difficult to elucidate the carcinogenic mechanism in the oral cavity, where various strains of bacteria coexist. The development of next-generation sequencers in recent years has enabled the performance of metaanalyses to investigate the composition of the entire bacterial flora, and many articles in the literature have reported the relationship between the disruption of homeostasis in bacterial flora (dysbiosis) and disease. Colorectal cancer, pancreatic cancer, esophageal cancer, oral cancer, and so on have been suggested to be associated with and correlated with specific bacteria, especially *Fusobacterium* species [27]. It has been found with oral cancer that the more orally indigenous bacteria *Corynebacterium* and *Kingella* there are, the less prominent the risk for head and neck cancer is [28]. However, it is difficult to study and analyze the effects of a massive bacterial flora on the host, and it is currently not clear how exactly the bacterial flora are involved in carcinogenesis. Interestingly, in the field, it has become widely accepted in recent years that intestinal bacterial flora have a significant impact on the local immune system, and in the field of tumors, there are reports of it being associated with cancer immunity, especially the immune checkpoint mechanism. A paper has also shown that the colonization of *Klebsiella* in the gastrointestinal tract might be involved in the development of inflammatory bowel disease by using a sterile mouse model, and research using sterile mouse models is also attracting attention in cancer research [29].

6.6 Candida

Like bacteria, fungi such as *Candida* have also been reported to be associated with oral cancer. According to the WHO, it is listed as one of the potentially malignant oral disorders. *Candida* is considered to be one of the causes of precancerous leukoplakia, but there are many points about its carcinogenic mechanism that are unclear. Meanwhile, *Candida* infection has a high incidence during radiochemotherapy in cancer treatment, and it is one of the well-known pathological conditions occurring as an adverse event during treatment.

7. Genetic mutation and signal transduction

Oral cancer is known to be caused by the accumulation of gene mutations. Mutations in these genes are transformed into traits that characterize individual cancers. Hanahan and Weinberg called these the hallmarks of cancer. In the year 2000, they formulated the six biological characteristics of sustaining proliferative signaling, evading growth suppressors, activating invasion and metastasis, enabling replicative immortality, inducing angiogenesis, and resisting cell death; in 2011, they added four more, bringing the total number to 10: avoiding immune destruction, tumor-promoting inflammation, genome instability and mutation, and deregulating cellular energetics. Genetic abnormalities are very common causes for the 10 cancer features [30] (Fig. 3). For genetic abnormalities in the head and neck region, there was a report in 2015 that comprehensively searched the genome of head and neck squamous epithelial cancer [31]. A search of 279 patients with head and neck squamous cell carcinomas revealed that there were four significant gene mutations in HPV-negative head and neck

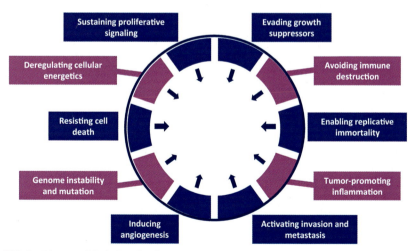

FIG. 3 Diagram of the hallmarks of cancer proposed by Hanahan and Weinberg. Six features were presented in 2000 (*blue*) and four were added in 2011 (*pink*). *Source: Hanahan D, Weinberg RA. Hallmarks of cancer: the next generation. Cell. 2011;144(5):646–674.*

10 Inflammation and oral cancer

squamous cell carcinoma: TP53, CDKN2A (p16), FAT1, and AJUBA; they mainly caused the loss of function of the tumor suppressor gene. Additionally, evaluations of signal activation showed that loss mutations of the functions of tumor suppressor genes, such as TP53 and CDKN2A (p16), were as high as 84% and 57%, respectively. However, oncogene hyperfunction mutations were few and were only up to approximately 12%, with EGFR, MYC, and HRAS at 15%, 14%, and 5%, respectively (Table 3). Furthermore, unlike lung cancer, there was no well-known driver mutation. It was unlikely that carcinogenesis would progress solely by disrupting the tumor suppressor gene. What are the drivers of head and neck squamous cell carcinoma? Genetic abnormalities result in the activation of signaling pathways that characterize cancer, such as cell proliferation. These have been attracting attention as therapeutic targets, as well as being characteristics of cancer.

TABLE 3 Alteration events for key genes are displayed by 279 patients with head and neck squamous cell carcinomas.

Candidate therapeutic targets and driver oncogenic events		
Receptor tyrosine kinases	HPV(−) $n = 243$	HPV(+) $n = 36$
EGFR	15%	6%
FGFR1	10%	0%
ERBB2	5%	3%
IGF1R	4%	0%
EPHA2	4%	3%
DDR2	3%	6%
FGFR2	2%	0%
FGFR3	2%	11%
MET	2%	0%
Oncogene		
CCND1	31%	3%
MYC	14%	3%
HRAS	5%	0%
Phosphoinositide 3-kinase		
PIK3CA	34%	56%
PTEN	12%	6%
PIK3R1	1%	3%
Tumor suppressor gene		
NF1	3%	0%
TP53	84%	3%
CDKN2A	58%	0%

Source: Cancer Genome Atlas N. Comprehensive genomic characterization of head and neck squamous cell carcinomas. Nature 2015;517(7536):576–582.

Epidemiology of oral cancer and its relationship with inflammation **Chapter | 1** **11**

7.1 EGFR pathway

The epidermal growth factor receptor (EGFR) is a receptor-type tyrosine kinase consisting of ErbB-1 (HER1/EGFR), ErbB-2 (HER2/Neu), ErbB-3(HER3), and ErbB-4 (HER4), and it is expressed in epidermal cells. Pathways, such as the RAS and MAPK pathway, PI3K and AKT pathway, and the Jak and STAT pathway, are induced downstream of the tyrosine kinase pathway, causing cell proliferation, infiltration, antiapoptosis, and angiogenesis in cancer cells. EGFR is overexpressed in various cancers, including 36%–100% of head and neck cancer, 40%–80% of lung cancer, 14%–91% of breast cancer, 33%–74% of gastric cancer, 25%–77% of colon cancer, and 50%–90% of kidney cancer [32]. Cetuximab is a treatment for oral cancer that targets these molecules. Cetuximab is an EGFR monoclonal antibody that is also used in colorectal cancer. Gefitinib, erlotinib, and lapatinib are also found as EGFR tyrosine kinase inhibitors.

7.2 RAS and MAPK pathway

The RAS and MAPK pathway is one of the most classic signaling pathways. It consists of three components, namely: H-RAS, K-RAS, and N-RAS. When RAS is activated, kinases activate a cascade of MAPKKK (Raf), MAPKK (MEK1/2), and MAPK (Erk). RAS mutations have been found in many forms of cancer, and the incidences are as follows when broken down by component: for K-RAS–pancreatic cancer, 97.7%; colorectal cancer, 44.7%; lung cancer, 30.9%; multiple myeloma, 22.8%; for N-RAS-malignant melanoma, 27.6%; multiple myeloma, 19.9%; thyroid cancer, 8.5%; colon cancer, 7.5%; for H-RAS-bladder cancer, 5.9%; head and neck cancer, 4.7%; thyroid cancer, 3.5%; and lung squamous epithelial carcinoma, 2.2%. Mutations in head and neck cancer are not as frequent, with K-RAS, N-RAS, and H-RAS occurring in 0.5%, 0.3%, and 4.7% of cases, respectively. H-RAS is frequently found in oral cancer associated with betel quid [33].

7.3 PI3K, AKT, and mTOR pathway

The PI3K and AKT pathway is one of the major signaling pathways in the EGFR tyrosine kinase pathway, and P13K is involved in the production of the second messenger, PIP3. Activation of PIP3 causes activation of the serine/threonine kinase AKT. mTOR is a serine/threonine kinase found downstream of AKT and is considered a major therapeutic target for this pathway. Everolimus, temsirolimus, and sirolimus are currently used as inhibitors of mTOR activation. The incidence of mutations in mTOR for cancer were 10% for endometrial cancer, 8% for gastric cancer, 8% for kidney cancer, 6% for skin cancer, and 5% for lung cancer. The incidence of mutations in head and neck cancer is approximately 3%, although abnormalities and increased expression are observed in many cases of head and neck cancer when the mutations and overexpression of upstream EGFR, PI3K, and AKT are combined [34].

7.4 NF-κB pathway

Nuclear factor-kappa B (NF-κB) is a signal transduction pathway consisting of five types of NF-κB families: p50, p52, p65, c-Rel, and RelB. NF-κB is inactivated by binding to IκB, which is normally present in the cytoplasm. The activation of NF-κB requires a kinase called IKK to phosphorylate IκB. The phosphorylated IκB is ubiquitinated and degraded, and the activated NF-κB translocates into the nucleus and regulates gene expression. The NF-κB pathway is a major regulator of the immune response and is activated by cytokines such as TNF-α and IL-1. As a result, it is also involved in the expression of many inflammatory cytokines such as TNF-α, IL-1, IL-6, and IL-8. The activation of these pathways is involved in crosstalk between NF-κB signaling in immune cells and cancer cells, acting in a way to promote tumor development [35]. It has also been reported that oral cancer is also involved in proliferation, angiogenesis/lymphangiogenesis, and infiltration/metastasis.

7.5 Jak and STAT pathway

The Jak and STAT pathway is a signaling pathway for the tyrosine kinase pathway in cytokine receptors such as EGFR and IL. Signals from the receptors lead to JAK activation and downstream STAT activation. STAT is composed of seven STAT transcription factor families, of which STAT1, 3, and 5 have been shown to be involved in the development and infiltration of head and neck cancer, and pSTAT3 expression has been associated with poor prognosis in oral squamous cell carcinoma. Clinical trials have been conducted in recent years on the JAK inhibitor Ruxolitinib for head and neck cancer [36].

7.6 WNT pathway

WNT signaling is an essential pathway during development in the embryonic period and controls the formation of various organs. It is an important signaling pathway in many types of cancer. WNT ligands cause disruption of the β-catenin degradation mechanism, which is composed of AXIN2, GSK3β, APC, and others. As a result, β-catenin accumulates in cells and promotes downstream gene expression, which in turn is said to affect cell proliferation, stem cell phenotype, EMT, and the tumor microenvironment in head and neck cancer. WNT signaling is also said to crosstalk with other signaling pathways. These crosstalk pathways include the TGFβ, NOTCH, and the Hippo pathways [37].

7.7 NOTCH pathway

NOTCH signaling is a well-conserved pathway in multicellular organisms that determines the fate of cells during development and maintains adult tissue homeostasis. NOTCH signaling is said to be carcinogenic or act in a

Epidemiology of oral cancer and its relationship with inflammation **Chapter | 1** **13**

tumor-suppressing manner in many types of cancer as a result of five ligands (JAGGED1 and 2; DLL1, 3, and 4) binding to cell surface ligands (NOTCH1, 2, 3, 4). Increased NOTCH signaling has been shown to correlate with poor prognosis in oral cancer, which suggests that NOTCH signaling might play a role in the early stages of oral cancer [38]. NOTCH is a cell surface receptor that transmits intercellular or short-range signals by interacting with transmembrane ligands on adjacent cells. The transmembrane domain must be cleaved and translocated into the nucleus by the γ-secretase in order for this signaling to work. There is ongoing research on the γ-secretase inhibitor, with a focus on this mechanism, for Alzheimer's disease, and it is also attracting attention as a cancer therapeutic target [39].

7.8 Hippo pathway

The Hippo pathway was discovered to be a pathway that controls the size of our bodies and organs. Activation of the Hippo pathway causes phosphorylation of the co-activators YAP and TAZ and suppression of gene transcription. It has been found in recent years that YAP activation due to Hippo pathway dysfunction leads to carcinogenesis, tumor progression, and chemotherapy resistance in many forms of cancer, such as lung cancer, colon cancer, ovarian cancer, and liver cancer [40]. YAP also functions as an oncogene in oral cancer, and it has been suggested that YAP overexpression in a mouse model may cause very early-stage cancer, and histopathologically increased YAP expression could be a risk factor for poor prognosis. Although the number of cases is small, this is expected to be a new marker for oral cancer [41].

7.9 PD-1/PD-L1 pathway

New drugs called "immune checkpoint inhibitors" have appeared on the scene in recent years and have been attracting attention in cancer treatment. The immune checkpoint mechanism is primarily a mechanism for controlling an excessive autoimmune response, although it has been found that some cancer cells use this to escape the immune response. Drugs targeted for the three checkpoint mechanisms PD-1, PD-L1, and CTLA-4, have already been developed. Two large-scale clinical trials have already been conducted on the anti-PD-1 antibodies Nivolumab and Pembrolizumab in head and neck cancer. The median overall survival was 7.5 months in the Nivolumab-treated group, as opposed to 5.1 months in the standard-treatment group [42]. For Pembrolizumab, with a combined positive score (CPS) of 20 or higher, the median overall survival for the combined Pembrolizumab + chemotherapy group was 13.0 months, as opposed to 10.7 months in the Cetuximab + chemotherapy group, with both prolonging survival significantly [43]. PD-1 antibody drugs are currently the only immune checkpoint inhibitors that can be used in oral cancer, and further expansion of their indications is desired.

14 Inflammation and oral cancer

8. Inflammation and cancer

Inflammation is a reaction for maintaining homeostasis in a living body when it is stimulated or invaded. It is a necessary defense mechanism for the living body, and inflammation disappears in most cases. However, inflammation may persist for a long period of time, depending on the type of irritation and invasion. A condition called chronic inflammation can cause autoimmune diseases, inflammatory diseases, cancer, and other conditions that are harmful to the individual. It is said that 15%–20% of human cancers are associated with chronic inflammation. The major causes of inflammation include biological, physical, and chemical factors. Biological factors include infectious diseases in general; bacteria, fungi, and viruses. Physical factors include mechanical stimuli, heat, ultraviolet rays, and radiation. Chemical factors include metals and organic compounds. Many of these are consistent with the risk factors for cancer development mentioned earlier.

Infectious diseases such as *Helicobacter pylori*-induced gastritis and gastric cancer, and hepatitis and liver cancer caused by the hepatitis virus, have been clarified in human cancers. In addition to these infectious diseases, mesothelioma caused by asbestos, as well as ulcerative colitis and colorectal cancer, have also been seen, although these are limited to specific types of cancer. Various hypotheses have been reported, such as genomic instability, oncogene activation, or suppressor gene dysfunction due to chronic inflammation. However, the mechanism by which inflammation contributes to tumorigenesis has not been elucidated. The same also applies to oral cancer, where chronic inflammation of the oral mucosa is common. In contrast, since cancer of the oral mucosa is relatively rare, the inflammatory process alone causing the development of cancer, in this case, is thought to be "rare" [44]. Cancer microenvironment and tumor immunity are fields that have attracted attention in recent years due to the progress in research on cancer and inflammation.

8.1 Cancer microenvironment

Cancer involves the formation of a network between cancer cells, cancer-associated fibroblasts (CAF), and blood vessels/lymphatic vessels, as well as the T cells, B cells, macrophages, dendritic cells, and others that migrate through these. It has become clear that within these environments, these networks work to develop and promote tumors by activating certain transcription factors and sometimes by promoting the expression of common signals. For example, fibroblasts that are observed during wound healing in normal tissue remain active around cancer cells. These fibroblasts are called cancer-associated fibroblasts (CAF), and it has been suggested that they are closely related, not only with cancer proliferation, but also angiogenesis through VEGF and infiltration and metastasis by MMP and TGF-β [45]. Furthermore, macrophages are one of the cell types that make up most of the cancer microenvironment and are called tumor-associated macrophages (TAM). Although macrophages play a role in

Epidemiology of oral cancer and its relationship with inflammation **Chapter | 1   15**

the immune response in normal tissues, they act instead to promote tumors around the cancer cells. Macrophages include M1 macrophages that have anti-bacterial, antiviral, and antitumor behaviors, and M2 macrophages that have tissue repair, immunosuppression, and tumor promotion behaviors. TAMs take the phenotype of M2 macrophages due to the influence of various cytokines. Many TAMs have been suggested to work in a manner similar to that of CAF, i.e., they are involved in angiogenesis by VEGF as well as infiltration/metastasis by MMP and TGF-β. These networks are also found in squamous cell carcinomas of the head and neck and are considered an exacerbating prognostic factor [46].

8.2   Cancer microenvironment and immune checkpoint mechanism

In the cancer microenvironment, even cells that are the main components of immunity (e.g., T cells and B cells) respond characteristically depending on the cytokine network within that microenvironment. In other words, the cancer microenvironment exerts a tumor-promoting effect by regulating the antitumor effect of B cells and acts to suppress the functions of CD8-positive T cells and CD4-positive T cells. Changes in the activity of these immune-responsive cells show a stronger inhibitory effect on cancer immunity due to the action of cells such as myeloid-derived suppressor cells (MDSC) and regulatory T cells in addition to TAMs. It has also become clear in recent years that PD-1 and CTLA4 are expressed on T cells, and research on the immune checkpoint mechanism among inflammatory cells is progressing. Interestingly, this mechanism not only shows antitumor activity in the network between cancer and tumor cells, but also shows antitumor activity by activating PD-1 and CTLA4 between antigen-presenting cells and T cells [47]. The expression of PD-1/PD-L1 (PD-L2) is also considered to be an exacerbating prognostic factor in the field of oral cancer. This has also been used in actual cancer treatment, such as by quantifying this expression through CPS (i.e., by comprehensively counting inflammatory cells as well as tumor cells) to predict the effects of immune checkpoint inhibitors.

References

[1] Bryant F. Cancer of the mouth. Boston Med Surg J 1919;181(15):452–8.

[2] Balkwill F, Mantovani A. Inflammation and cancer: back to Virchow? Lancet 2001;357(9255):539–45.

[3] Tanaka T, Ishigamori R. Understanding carcinogenesis for fighting oral cancer. J Oncol 2011;2011:603740.

[4] Sung H, Ferlay J, Siegel RL, Laversanne M, Soerjomataram I, Jemal A, et al. Global cancer statistics 2020: GLOBOCAN estimates of incidence and mortality worldwide for 36 cancers in 185 countries. CA Cancer J Clin 2021.

[5] Japan Society for Head and Neck Cancer CRC. Report of Head and Neck Cancer Registry of Japan Clinical Statistics of Registered Patients; 2017. Available from: http://www.jshnc.umin. ne.jp/pdf/HNCreport_2017.pdf.

16 Inflammation and oral cancer

[6] Japan Society for Head and Neck Cancer CRC. Report of Head and Neck Cancer Registry of Japan Clinical Statistics of Registered Patients; 2012. Available from: http://www.jshnc.umin. ne.jp/pdf/HNCreport_2012.pdf.

[7] Japan Society for Head and Neck Cancer CRC. Report of Head and Neck Cancer Registry of Japan Clinical Statistics of Registered Patients; 2016. Available from: http://www.jshnc.umin. ne.jp/pdf/HNCreport_2016.pdf.

[8] Japan Society for Head and Neck Cancer CRC. Report of Head and Neck Cancer Registry of Japan Clinical Statistics of Registered Patients; 2011. Available from: http://www.jshnc.umin. ne.jp/pdf/HNCreport_2011.pdf.

[9] Japan Society for Head and Neck Cancer CRC. Report of Head and Neck Cancer Registry of Japan Clinical Statistics of Registered Patients; 2013. Available from: http://www.jshnc.umin. ne.jp/pdf/HNCreport_2013.pdf.

[10] Japan Society for Head and Neck Cancer CRC. Report of Head and Neck Cancer Registry of Japan Clinical Statistics of Registered Patients; 2014. Available from: http://www.jshnc.umin. ne.jp/pdf/HNCreport_2014.pdf.

[11] Japan Society for Head and Neck Cancer CRC. Report of Head and Neck Cancer Registry of Japan Clinical Statistics of Registered Patients; 2015. Available from: http://www.jshnc.umin. ne.jp/pdf/HNCreport_2015.pdf.

[12] Ferlay J, Soerjomataram I, Dikshit R, Eser S, Mathers C, Rebelo M, et al. Cancer incidence and mortality worldwide: sources, methods and major patterns in GLOBOCAN 2012. Int J Cancer 2015;136(5):E359–86.

[13] Ferlay J, Shin HR, Bray F, Forman D, Mathers C, Parkin DM. Estimates of worldwide burden of cancer in 2008: GLOBOCAN 2008. Int J Cancer 2010;127(12):2893–917.

[14] Parkin DM, Bray F, Ferlay J, Pisani P. Estimating the world cancer burden: Globocan 2000. Int J Cancer 2001;94(2):153–6.

[15] Parkin DM, Bray F, Ferlay J, Pisani P. Global cancer statistics, 2002. CA Cancer J Clin 2005;55(2):74–108.

[16] Bray F, Ferlay J, Soerjomataram I, Siegel RL, Torre LA, Jemal A. Global cancer statistics 2018: GLOBOCAN estimates of incidence and mortality worldwide for 36 cancers in 185 countries. CA Cancer J Clin 2018;68(6):394–424.

[17] Lin WJ, Jiang RS, Wu SH, Chen FJ, Liu SA. Smoking, alcohol, and betel quid and oral cancer: a prospective cohort study. J Oncol 2011;2011:525976.

[18] Wyss A, Hashibe M, Chuang SC, Lee YC, Zhang ZF, Yu GP, et al. Cigarette, cigar, and pipe smoking and the risk of head and neck cancers: pooled analysis in the International Head and Neck Cancer Epidemiology Consortium. Am J Epidemiol 2013;178(5):679–90.

[19] Hashibe M, Brennan P, Benhamou S, Castellsague X, Chen C, Curado MP, et al. Alcohol drinking in never users of tobacco, cigarette smoking in never drinkers, and the risk of head and neck cancer: pooled analysis in the International Head and Neck Cancer Epidemiology Consortium. J Natl Cancer Inst 2007;99(10):777–89.

[20] Prabhu RV, Prabhu V, Chatra L, Shenai P, Suvarna N, Dandekeri S. Areca nut and its role in oral submucous fibrosis. J Clin Exp Dent 2014;6(5):e569–75.

[21] Durst M, Gissmann L, Ikenberg H, zur Hausen H. A papillomavirus DNA from a cervical carcinoma and its prevalence in cancer biopsy samples from different geographic regions. Proc Natl Acad Sci U S A 1983;80(12):3812–5.

[22] Nordfors C, Vlastos A, Du J, Ahrlund-Richter A, Tertipis N, Grun N, et al. Human papillomavirus prevalence is high in oral samples of patients with tonsillar and base of tongue cancer. Oral Oncol 2014;50(5):491–7.

Epidemiology of oral cancer and its relationship with inflammation **Chapter | 1 17**

[23] Kobayashi K, Hisamatsu K, Suzui N, Hara A, Tomita H, Miyazaki T. A review of HPV-related head and neck cancer. J Clin Med 2018;7(9).

[24] Ad S, Jr SF, Klein G, Henle W, Henle G, De-Thé G, et al. Epstein–Barr virus-associated antibody patterns in carcinoma of the post-nasal space.pdf. Clin Exp Immunol 1969;5(5):443–59.

[25] Ayee R, Ofori MEO, Wright E, Quaye O. Epstein Barr virus associated lymphomas and epithelia cancers in humans. J Cancer 2020;11(7):1737–50.

[26] Rytkonen AE, Hirvikoski PP, Salo TA. Lymphoepithelial carcinoma: two case reports and a systematic review of oral and sinonasal cases. Head Neck Pathol 2011;5(4):327–34.

[27] Fujiwara N, Kitamura N, Yoshida K, Yamamoto T, Ozaki K, Kudo Y. Involvement of fusobacterium species in oral cancer progression: a literature review including other types of cancer. Int J Mol Sci 2020;21(17).

[28] Hayes RB, Ahn J, Fan X, Peters BA, Ma Y, Yang L, et al. Association of oral microbiome with risk for incident head and neck squamous cell cancer. JAMA Oncol 2018;4(3):358–65.

[29] Atarashi K, Suda W, Luo C, Kawaguchi T, Motoo I, Narushima S, et al. Ectopic colonization of oral bacteria in the intestine drives TH1 cell induction and inflammation. Science 2017;358(6361):359–65.

[30] Hanahan D, Weinberg RA. Hallmarks of cancer: the next generation. Cell 2011;144(5):646–74.

[31] Cancer Genome Atlas N. Comprehensive genomic characterization of head and neck squamous cell carcinomas. Nature 2015;517(7536):576–82.

[32] Normanno N, Maiello MR, De Luca A. Epidermal growth factor receptor tyrosine kinase inhibitors (EGFR-TKIs): simple drugs with a complex mechanism of action? J Cell Physiol 2003;194(1):13–9.

[33] Cox AD, Fesik SW, Kimmelman AC, Luo J, Der CJ. Drugging the undruggable RAS: mission possible? Nat Rev Drug Discov 2014;13(11):828–51.

[34] Polivka Jr J, Janku F. Molecular targets for cancer therapy in the PI3K/AKT/mTOR pathway. Pharmacol Ther 2014;142(2):164–75.

[35] Taniguchi K, Karin M. NF-kappaB, inflammation, immunity and cancer: coming of age. Nat Rev Immunol 2018;18(5):309–24.

[36] Geiger JL, Grandis JR, Bauman JE. The STAT3 pathway as a therapeutic target in head and neck cancer: barriers and innovations. Oral Oncol 2016;56:84–92.

[37] Alamoud KA, Kukuruzinska MA. Emerging insights into Wnt/beta-catenin signaling in head and neck cancer. J Dent Res 2018;97(6):665–73.

[38] Hatano K, Saigo C, Kito Y, Shibata T, Takeuchi T. Overexpression of JAG2 is related to poor outcomes in oral squamous cell carcinoma. Clin Exp Dent Res 2020;6(2):174–80.

[39] Nowell CS, Radtke F. Notch as a tumour suppressor. Nat Rev Cancer 2017;17(3):145–59.

[40] Harvey KF, Zhang X, Thomas DM. The hippo pathway and human cancer. Nat Rev Cancer 2013;13(4):246–57.

[41] Ono S, Nakano K, Takabatake K, Kawai H, Nagatsuka H. Immunohistochemistry of YAP and dNp63 and survival analysis of patients bearing precancerous lesion and oral squamous cell carcinoma. Int J Med Sci 2019;16(5):766–73.

[42] Ferris RL, Blumenschein Jr G, Fayette J, Guigay J, Colevas AD, Licitra L, et al. Nivolumab for recurrent squamous-cell carcinoma of the head and neck. N Engl J Med 2016;375(19):1856–67.

[43] Burtness B, Harrington KJ, Greil R, Soulières D, Tahara M, de Castro G, et al. Pembrolizumab alone or with chemotherapy versus cetuximab with chemotherapy for recurrent or metastatic squamous cell carcinoma of the head and neck (KEYNOTE-048): a randomised, open-label, phase 3 study. Lancet 2019;394(10212):1915–28.

18 Inflammation and oral cancer

[44] Feller L, Altini M, Lemmer J. Inflammation in the context of oral cancer. Oral Oncol 2013;49(9):887–92.

[45] Ziani L, Chouaib S, Thiery J. Alteration of the antitumor immune response by cancer-associated fibroblasts. Front Immunol 2018;9:414.

[46] Noy R, Pollard JW. Tumor-associated macrophages: from mechanisms to therapy. Immunity 2014;41(1):49–61.

[47] Ribas A. Tumor immunotherapy directed at PD-1. N Engl J Med 2012;366(26):2517–9.

Chapter 2

Diagnosis of oral squamous cell carcinomas and precancerous lesions

Keisuke Nakano and Hitoshi Nagatsuka

Department of Oral Pathology and Medicine, Graduate School of Medicine, Dentistry and Pharmaceutical Sciences, Okayama University, Okayama, Japan

Abbreviations

OPMDs oral potentially malignant disorders
OED   oral epithelial dysplasia
OSCC   oral squamous cell carcinoma
PVL   proliferative verrucous leukoplakia

1. Introduction

The annual number of head and neck cancer cases is approximately 350,000 worldwide, and this type of cancer accounts for approximately 2.0% of all cancers [1]. Oral cancer occurs in 4 per 100,000 individuals and is more commonly seen in men, with a male-to-female sex ratio of approximately 2:1. However, in low-income countries, the proportion of female patients is higher. The areas with a high morbidity predominantly include South Asia, Southeast Asia, and Pacific Rim countries with populations with a habit of using smokeless tobacco (betel nut chewing) [2]. It is estimated that the overall prevalence of head and neck cancer worldwide will increase by 62% by 2035 [3]. Scientific evidence has shown that smoking, smokeless tobacco use (betel nut chewing), excessive drinking, and human papillomavirus (HPV) infection can be used to estimate the causal relationship between the development of oral and pharyngeal cancers [4]. These risk factors are also considered to be the cause of the prevalence of oral potentially malignant disorders (OPMDs). In the comparison of the changes in morbidity between oral cancer and laryngeal cancer over the years, there is no change observed in oral cancer in both men and women. However, in men, the prevalence of laryngeal cancer has increased in recent years. This is associated with an increase in the prevalence of HPV infection. In North America, the prevalence of laryngeal cancer is increasing, while that of oral

Inflammation and Oral Cancer. https://doi.org/10.1016/B978-0-323-88526-3.00002-6
Copyright © 2022 Elsevier Inc. All rights reserved.

20 Inflammation and oral cancer

cancer is declining. It has been pointed out that this is related to the decrease in the smoking rate. The development of oral cancer is also associated with oral hygiene and nutritional status.

2. OPMDs

2.1 Definition and classification of OPMDs

In the WHO Classification of Head and Neck Tumors 4th edition, OPMDs are defined as follows: clinical presentations that carry a risk of cancer development in the oral cavity, whether in a clinically definable precursor lesion or clinically normal oral mucosa [5]. Table 1 shows the list of OPMDs.

OPMDs are caused by a variety of factors. These include smoking and alcohol consumption; however, in many cases, the cause is not clear. The following is a description of typical OPMDs.

2.2 Leukoplakia

Leukoplakia is both a general term for lesions showing white plaque-like or patchy ridges on the oral mucosa, and a clinical diagnosis term. It is more common in elderly populations, predominantly occurring between the ages of 50 and 70 years, and in men. Leukoplakia is often found on the buccal mucosa, tongue, gums, floor of the mouth, and oral palate. Its etiology is thought to involve locally acting physical, chemical, and biological stimuli, such as smoking, alcohol consumption, incompatible prostheses, and Candida infection. The canceration

TABLE 1 Oral potentially malignant disorders.

Erythroplakia
Erythroleukoplakia
Leukoplakia
Oral submucous fibrosis
Dyskeratosis congenital
Smokeless tobacco keratosis
Palatal lesions associated with reverse smoking
Chronic candidiasis
Lichen planus
Discoid lupus erythematosus
Syphilitic glossites
Actinic keratosis (lip only)

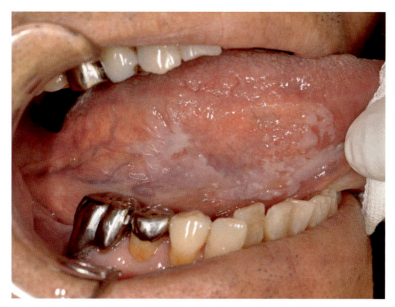

FIG. 1 Leukoplakia. A slightly raised white lesion is seen from the edge of the tongue to the underside of the tongue.

rate of leukoplakia is 4%–18%. Clinically, wart-like or tumor-like lesions that occur on the rim of the tongue, lower surface of the tongue, and floor of the mouth, or those with ulcers, are believed to have a high risk of becoming cancerous. Histopathologically, increased keratinization and oral epithelial dysplasia (OED) are common; however, there are also cases of carcinoma in situ (CIS) and squamous cell carcinoma (SCC) (Fig. 1).

2.3 Erythroplakia

Erythroplakia is a lesion that appears as velvety erythema on the oral mucosa. Although its incidence is lower than that of leukoplakia, the canceration rate is high, at 40%–50%. The predominant age at which it occurs is from 50 to 60 years, and there is no difference by sex. It often occurs on the buccal mucosa, tongue, oral palate, floor of the mouth, and gums, and is often accompanied by contact pain.

Histopathologically, erythroplakia is often accompanied by marked epithelial atrophy and atypia. In some cases, histologically, it should be regarded as OED or early-stage SCC (Fig. 2).

2.4 Oral lichen planus

Oral lichen planus is a chronic inflammatory disease with abnormal keratinization, and linear and reticulated vitiligo is observed on the oral mucosa with erythema. The buccal mucosa is the most common site, and it is often found bilaterally. It is reported that canceration occurs in 2%–3% of all cases. Vitamin

22 Inflammation and oral cancer

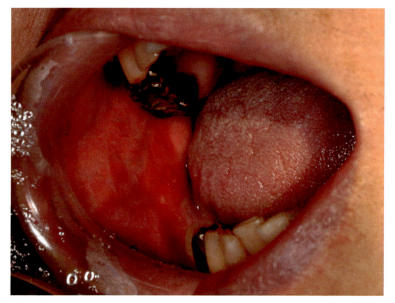

FIG. 2 Erythroplakia. A bright red lesion is seen on the buccal mucosa.

deficiency, metal allergies, and viral infections are the likely causes, and the involvement of immune disorders is regarded as important.

Histopathologically, oral lichen planus is characterized by hyperkeratinization, irregular thickening of the spinal cell layer, damage to basal cells, and banded infiltration of lymphocytes in the subepithelial connective tissue (Fig. 3A and B).

2.5 Oral submucous fibrosis

Oral submucous fibrosis often occurs in Southeast Asian regions with populations that have a habit of betel nut chewing. Yellowish white lesions of the mucosa are found on the palate and buccal mucosa, and the mucosa in the area loses elasticity and becomes hard. It is reported that approximately 7% of cases become cancerous. In Southeast Asia, oral cancer accounts for approximately 30% of all cancer types, and its relevance has been pointed out.

Histologically, atrophy of the oral mucosal epithelium and various degrees of epithelial dysplasia are observed. Subepithelial connective tissue is poorly vascularized, with vitrification, chronic inflammatory cell infiltration, and fibrotic changes (Fig. 4).

2.6 Oral candidiasis

Oral candidiasis is an infection of the oral mucosa with *Candida albicans*. Candida infection is an opportunistic infection that is thought to be caused by a decline in the host immune system, and the development of chronic oral

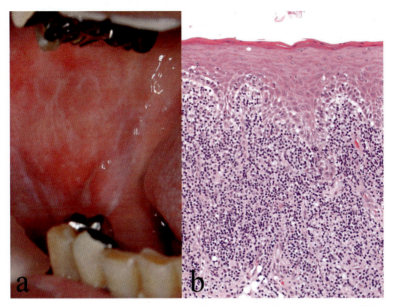

FIG. 3 Oral lichen planus. (A) Reticulated vitiligo is found on the buccal mucosa. (B) In the epithelial tissue, hyperkeratinization, elongation of epithelial legs, damage to basal cells, and lymphocyte infiltration into subepithelial tissue are observed.

FIG. 4 Oral submucous fibrosis. Infiltration of inflammatory cells and increased fibrosis are observed under the epithelial tissue.

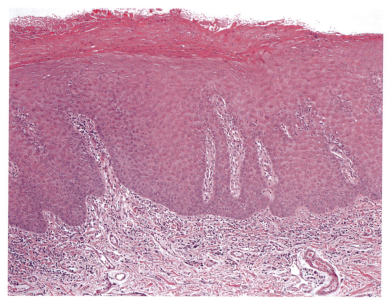

FIG. 5 Oral candidiasis. Invasion of Candida can be seen in the epithelial tissue.

mucositis is thought to be involved in the development of oral cancer. It often occurs in elderly and immunocompromised patients and is most often seen in the tongue, buccal mucosa, and palate. The lesion appears grossly white; however, an erythema type is also present. Histopathologically, invasion of *Candida* into epithelial tissue is observed (Fig. 5).

2.7 Oral syphilis

Oral syphilis is an infection with *Treponema pallidum*. Syphilis is divided into congenital and acquired, according to the infection route; however, acquired syphilis follows a process that is classified into four stages. In the first stage, early induration and chancre (ulcer associated with infection) are found in the early infected area. In the second stage, leukoplakia-like lesions, such as vitiligo and papular syphilis, are found on the tongue and soft palate, and the risk of canceration is high. In the third stage, granulomatous inflammation occurs in soft and bone tissues, and ulceration is often seen in the tongue and palate. In the oral cavity, Stage 3 lesions are more frequent than Stage 1 and Stage 2 lesions, and syphilitic glossitis is thought to be more likely to develop into SCC.

2.8 Discoid lupus erythematosus

Discoid lupus erythematosus is an irregular morphological erythematous lesion with keratinization and erosion that often occurs on the lips and buccal mucosa. It is considered to be an autoimmune disease, and the occurrence of carcinoma of the lip has mainly been reported.

2.9 Proliferative verrucous leukoplakia

Proliferative verrucous leukoplakia (PVL) is a distinct and aggressive type of OPMD. Unlike normal leukoplakia, it is a latent oral malignant disease that is multiple and often widespread and has a high rate of canceration and recurrence. Proliferative verrucous leukoplakia is often found in elderly women over the age of 60 years and is most common in the gums, alveolar mucosa, and palate. It has been reported that in the tongue, the development in the gingiva leads to cancer at a high rate. Regarding the etiology, the relationship with carcinogens of oral cancer, such as alcohol, tobacco, and HPV, is not clear. This disease has a progressive course, and the histology differs depending on the stage. In the early stage, the lesion is flat and localized, and the epithelium shows hyperkeratosis. As the stage progresses, the lesion spreads horizontally while gradually becoming verrucous. In the late stage, it transitions to SCC via OED.

3. Oral epithelial dysplasia

3.1 Definition and classification of OED

In the WHO classification 4th edition, OED is defined as "a spectrum of architectural and cytological epithelial changes caused by accumulation of genetic changes, associated with an increased risk of progression to SCC" [5] and is classified according to the degree of structural atypia and cell atypia, as shown in Table 2. It is classified as mild dysplasia when the atypical cells occupy 1/3 of all layers of epithelial tissue; severe dysplasia, 2/3 or more; and moderate dysplasia, in between (Fig. 6). In addition, the WHO classification 4th edition proposes two classification methods for OED (low grade and high grade), along with the conventional three classification methods. In OED, superficial dysplasia may be scarce despite marked cell atypia below the epithelium. The WHO

TABLE 2 Diagnostic criteria for oral epithelial dyspla.

Architectural changes	Cytologocal changes
Irregular epithelial stratification	Abnormal variation in nuclear size
Loss of polarity of basal cells	Abnormal variation in nuclear shape
Drop-shaped rete ridges	Abnormal variation in cell size
Increased number of mitotic figures	Abnormal variation in cell shape
Abnormally superficial mitotic figures	Increased N:C ratio
Premature keratinization in single cells	Atypical mitotic figures
Keratin pearls within rete ridges	Increased number and size of nucleoli
Loss of epithelial cell cohesion	Hyperchromasia

26 Inflammation and oral cancer

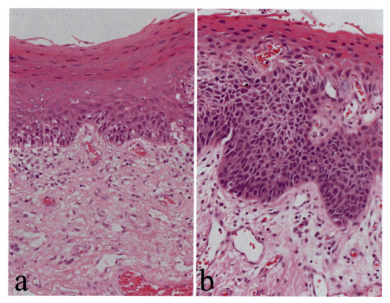

FIG. 6 Oral epithelial dysplasia. (A) *Mild dysplasia*: atypical cells and structural atypia are found in the lower layer of the epithelial layer. (B) *Severe dysplasia*: epithelial cells with strong atypia are found near the surface layer.

classification 4th edition allows diagnosis of severe dysplasia if the atypia is severe, even if the atypia is localized below the epithelial layer. Reactive atypical epithelium associated with inflammation and epithelial regeneration is often confused with epithelial dysplasia. However, structural atypia recognized by the appearance of regional atypical cells in epithelial tissue, dyskeratosis, and Ki-67-positive images can be helpful for diagnosis. Immunostaining of p53, Ki-67, and cytokeratin (CK13 and CK17) is useful as a diagnostic aid. Epithelial dysplasia caused by HPV infection is rare in the oral cavity, and the presence or absence of infection does not serve as a diagnostic indicator.

3.2 Immunohistochemistry (IHC) of OED

IHC is useful for diagnosing OED. As the oral mucosa is an environment that is susceptible to external stimuli, it often exhibits inflammation caused by various stimuli and atypia associated with regenerative repair. In OED, there is a tendency for regional atypia to appear in the epithelial tissue, and this differentiates it from reactive atypia. As immunohistochemical diagnostic markers, p53, Ki-67, and cytokeratin (CK13 and CK17) are used to aid in the diagnosis [3]. p53 is negative in normal mucosal epithelium and hyperplastic epithelium or shows a small number of positivity in the basal layer (Fig. 7A–E). In epithelial dysplasia and CIS, positivity increases in areas showing cellular and structural atypia (Fig. 8A–E). Ki-67 is positive in the nucleus of parabasal cells in normal

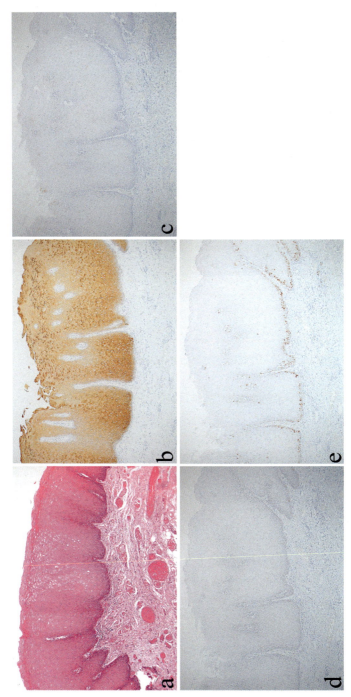

FIG. 7 Immunohistochemistry of epithelial hyperplasia. (A) In the hyperplastic epithelium, thickening of the epithelium, mainly in the spinous layer, is observed. (B) In the hyperplastic epithelium, the entire hyperplastic epithelium, from the basal to the surface layer, is CK13-positive. (C) The hyperplastic epithelium is CK17-negative. (D) Scattered p53-positive basal cells of the hyperplastic epithelium. (E) Ki-67-positive parabasal cells of the hyperplastic epithelium.

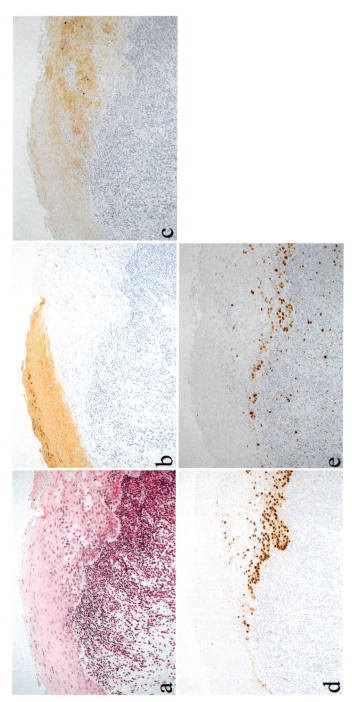

FIG. 8 Immunohistochemistry of oral epithelial dysplasia. (A) In oral epithelial dysplasia, the regional atypical epithelium appears in contact with normal epithelial tissue. (B) The normal epithelium is CK13-positive. (C) Atypical epithelial dysplasia of the oral epithelial dysplasia is CK17-positive. (D) Cells in the basal to middle layers of oral epithelial dysplasia are P53-positive. (E) Cells in the basal to middle layers of oral epithelial dysplasia are Ki-67-positive.

mucosa and hyperplastic epithelium (Fig. 7E). However, in OED and CIS, basal cells become positive, and increasing positivity is noted from the middle layer of stratified squamous epithelium to the superficial epithelial cells (Fig. 8E). The expression of Ki-67 in epithelial dysplasia is an indicator not only of cell proliferative activity, but also structural atypia of epithelial tissue. Cytokeratin is CK13-positive and CK17-negative in normal mucosa and hyperplastic epithelium (Fig. 7B and C). However, in OED and CIS, CK13 is lost, and CK17 turns positive (Fig. 8B and C). Changes in the expression of CK13 and CK17 may also be seen in reactive atypical epithelial tissues; thus, they should be used only as an aid to diagnosis.

4. Oral cancer

4.1 Epidemiology of oral cancer

More than 90% of oral malignancies are epithelial malignancies. Most of the carcinomas that occur in the oral cavity are SCCs derived from the mucosal epithelium of the oral cavity. Herein, SCC derived from the mucosa anatomically covering the oral cavity will be described as oral cancer. Epidemiological data on oral cancer can be found in the GLOBOCAN database published by the International Agency for Research on Cancer, an external organization of the WHO [1]. Oral cancer accounts for approximately 2% of all cancers, and the proportion of affected men is twice as high as that of affected women. Various causes and triggers can be considered for oral cancer. Mainly, these include smoking and alcohol consumption; however, chemical irritation from various foods can also be a cause of the disease. Smokeless tobacco use (betel nut chewing) is widespread in Southeast Asia and India and is one of the main causes of oral cancer. In addition, persistent chronic irritation due to caries, defective prostheses associated with dental treatment, incompatible dentures, and dentition irregularities also promote the development of oral cancer by synergistic action with other carcinogenic factors. HPV infection has been shown to cause cervical cancer; however, oral cancer may also be caused by HPV infection. In addition, it is known that bacterial infections, such as candidiasis and atrophic glossitis seen in the third stage of syphilis, have progressed from latent oral malignancies to cancer.

4.2 Clinical findings of oral cancer

Initially, oral cancer is seen as erosions and ulcers with mixed vitiligo and erythema and nodular or superficial granular elevated lesions. As it progresses, it shows various clinical presentations, such as formation of tumors with ulcers and induration. Depending on the site, tongue cancer tends to be associated with more pain and ulceration, while gingival cancer tends to be present as a painless mass.

4.3 Oral cancer histology

Approximately 90% of the histological types of oral cancer are SCC. The subtypes of SCC include basaloid SCC, spindle cell SCC, adenosquamous carcinoma, carcinoma cuniculatum, verrucous SCC, lymphoepithelial carcinoma, papillary SCC, and acantholytic SCC. In addition, adenocarcinoma derived from the salivary glands and undifferentiated cancer can develop.

4.4 Squamous cell carcinoma

SCC is a tumor involving the growth of tumor cells similar to stratified squamous epithelium. It exhibits infiltrative growth while forming vesicles of tumor cells with atypia. Its basic histology is basal cell-like tumor cells arranged on the margin of the tumor nests, causing layered differentiation of the squamous epithelium toward the center of the nests. Keratinization is observed in the central part. The stratum corneum found in the tumor nest is called a cancer pearl. Tumor cells tend to differentiate into stratified squamous epithelium but are classified into well-differentiated, moderately differentiated, and poorly differentiated, according to the degree of tumor differentiation (Fig. 9). Oral SCC mainly shows lymphoid metastasis. Tumor cells that infiltrate the lymph vessels

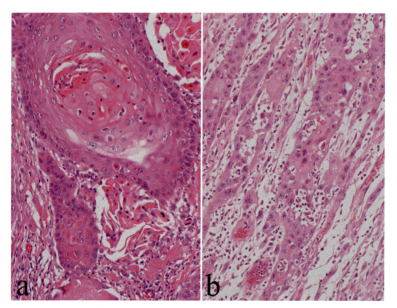

FIG. 9 Squamous cell carcinoma. (A) Well-differentiated squamous cell carcinoma. Tumor cells with a marked tendency to differentiate into squamous epithelium form tumor nests, and keratinous foci are formed in the center of the nests. (B) Poorly differentiated squamous cell cancer. Tumor cells have a poor tendency to differentiate into the squamous epithelium; the tumor nests are cord-like, and no keratinous foci are formed.

Diagnosis of oral squamous cell carcinomas and precancerous lesions Chapter | 2   31

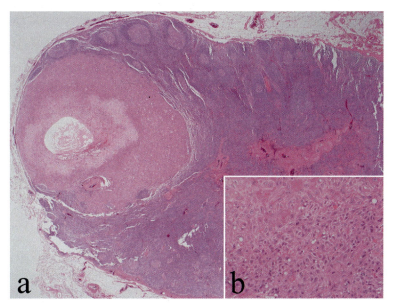

FIG. 10 Squamous cell carcinoma and lymph node metastasis. (A) A solid mass of tumor cells in the lymph nodes is noted. (B) Enlarged view of the tumor area. Proliferation of tumor cells showing a tendency to differentiate into squamous epithelium is observed.

in the tissue metastasize to local lymph nodes near the primary lesion (Fig. 10). Those prone to metastasis in the head and neck region include upper internal jugular vein lymph nodes and submandibular lymph nodes. The lungs and bone tissues are the organs that are prone to hematogenous distant metastasis. It is said that many carcinomas, including oral cancer, have a gene abnormality of p53, a tumor suppressor gene. Normally, the half-life of normal p53 protein is as short as a few minutes. However, the half-life of mutant p53 protein is long. As it accumulates in the cytoplasm, it can be stained immunohistochemically, the results of which are also used as an aid in cancer diagnosis [6].

4.5 Grade classification of SCC

The WHO classification presents the grade classification as a general histological malignancy evaluation method for SCC. This is derived from the Broders classification of lip cancer and is a classification based on the degree of differentiation into squamous epithelium. SCC is classified as G1 to G3. G1 (well-differentiated type) has a cellular differentiation and structure similar to that of normal stratified squamous epithelium. Tumor cells have a clear tendency to differentiate into squamous cells. The basal cell-like cells are arranged at the margin of the tumor nest and show stratified differentiation into spinous cell-like cells and keratinized cells toward the central part of the nest. Tumor

32   Inflammation and oral cancer

cells have intercellular bridges, with clear tendency of keratinization, and the formation of keratinocytes is observed. It is scarce in atypical mitotic figures and polynuclear cancer cells. The keratinization is prominent in the tumor nests and is observed in most of the infiltrating and proliferating tumor nests.

In G2 (moderately differentiated type), the tendency of differentiation into stratified squamous epithelium is diminished, and stratified differentiation becomes unclear. The number of keratinocytes is reduced, and the formation of clear keratinocytes is poor. Furthermore, the intercellular bridges become obscured. Compared to those in the well-differentiated type, the cell atypia becomes more obvious, and the number of cell division images accompanied by atypical nuclear division increases. Within a solid tumor lesion with a tendency to differentiate into squamous epithelium, partial keratinization is seen.

G3 (poorly differentiated type) shows poor differentiation into stratified squamous epithelium, and there is loss of stratified differentiation and keratinization tendency. Tumor cells that show strong cell atypia are the main constituents. There are also many cell division images with atypical nuclear division. In poorly differentiated squamous epithelial carcinoma, infiltrative and proliferating nodules or cord-like nests are formed, and keratinization is hardly observed in the nests.

4.6   IHC for SCC

Many oral SCCs have TP53 oncogene abnormalities; in this type of carcinoma, p53 is useful as a diagnostic marker. Most of the tumor cells of oral squamous cell carcinoma (OSCC) are immunohistochemically positive for p53, and many tumor cells labeled with Ki-67 are observed (Fig. 12). Although p63 is diffusely positive for many tumor cells in the tumor parenchyma, it is known that tumor cells for myoepithelial tumors and adenocarcinomas also show positivity. Therefore, it is necessary to recognize that it is not a marker specific to SCC. The expression pattern of p40 differs depending on the degree of tumor differentiation. In well-differentiated SCC, there is a high positivity among tumor cells at the tip of infiltration, while in poorly differentiated SCC, there is positivity in most of the tumor cells in the entire tumor nest [4]. In addition, OSCC has limited evidence of carcinogenesis due to HPV infection, and HPV-related factors have minimal utility as diagnostic markers.

4.7   CIS

CIS is a tumor in which cancer cells grow locally in the epithelium and do not destroy the basement membrane or infiltrate the subepithelial connective tissue. According to the conventional histological definition, the appearance of atypical cells extends to all layers of the epithelium with a marked loss of polarity in the epithelial tissue [7]. This has applied the definition of CIS of the cervix to that of the oral cavity. In actual clinical settings, there are few CIS of the oral mucosa that meet these diagnostic criteria. In the oral mucosa, although there are marked cell atypia and structural atypia on the basal cell side, there is a type of CIS in which well-differentiated cells are arranged on the epithelial surface layer. It is known to easily progress to invasive cancer. The incidence of well-differentiated SCC is high in the oral cavity. As such, it is likely that most of the CIS, considered to be the prestage of invasive carcinoma, maintains differentiation and maturation into stratified squamous epithelium. According to the Japanese Society of Oral Pathology, CIS with stratified squamous epithelium and a clear tendency to keratinize is classified as superficial differentiated CIS (Fig. 11A) and CIS with cellular atypia in all epithelial layers as basal cell epithelial carcinoma (Fig. 11B). Most of the CIS that appears in the oral mucosa is superficial differentiated CIS and requires caution. Regarding CIS of the oral cavity, there is no clear description as an independent disease in the WHO classification (2017). It is explained as being synonymous with severe dysplasia or that CIS is included in high-grade dysplasia [5]. However, it is necessary to diagnose CIS for those that are clearly histopathologically considered to be CIS. The Japanese Society of Oral Pathology presents various histological images of CIS in an "oral CIS catalog" [8]: (http://www.jsop.or.jp/wp/wp-content/themes/jsop_pc/images/oralCIScatalog_s.pdf).

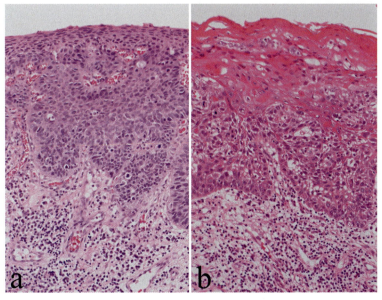

FIG. 11 Carcinoma in situ (JSOP). (A) CIS (JSOP), basaloid type. (B) CIS (JSOP), differentiated type.

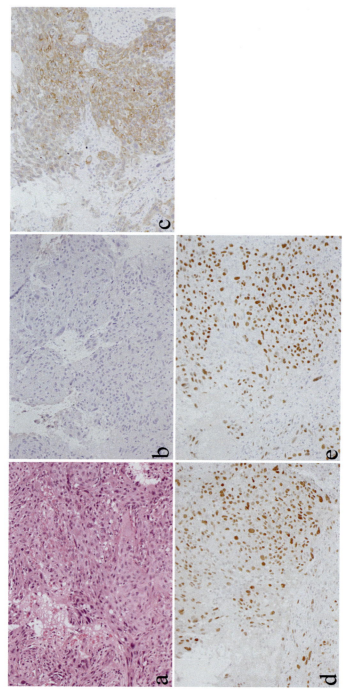

FIG. 12 Immunohistochemistry of oral squamous cell carcinoma. (A) In oral squamous epithelial carcinoma, atypical strong tumor cell proliferation that tends to differentiate into squamous epithelium is observed. (B) Tumor cells are CK13-negative. (C) Tumor cells are CK17-positive. (D) Most tumor cells are p53-positive. (E) Most tumor cells are Ki-67-positive.

5. Subtype of oral SCC

5.1 Basaloid SCC

Basaloid SCC is a highly malignant subtype of SCC involving proliferation of basal cell-like tumor cells. Basal cell-like cells form tumor nests that exhibit solid proliferation and are often accompanied by necrosis in the center of the nest [9, 10]. A palisade arrangement can be seen at the margin of the tumor nest. Many mitotic figures are found in the tumor nest. In rare cases, it may have a cribriform pattern and may need to be differentiated from other malignant salivary gland tumors and metastatic adenocarcinomas (Fig. 13). Basaloid SCC is often accompanied by deposition of PAS-stained positive basement membrane material. Occurrence is rare in the oral cavity but common in the pharynx; it is most often seen in men in their 60s to 80s. The prognosis is poor, with regional lymph node metastasis in approximately 60% and distant metastasis to the lungs and bones in 35%–50%.

5.2 Spindle cell SCC

Histologically, spindle cell SCC involves sarcoma-like growth of tumor cells that do not show keratinization—mainly spindle-shaped cells [11, 12]. Part of the tumor tissue may be associated with regular SCC. Tumor cells are rich in polymorphism, have strong nuclear atypia, and are immunohistochemically

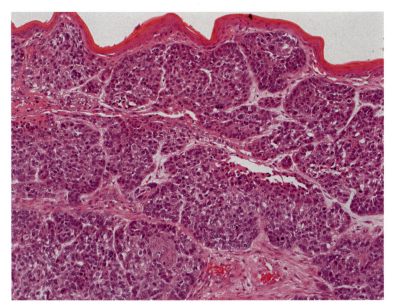

FIG. 13 Basaloid squamous cell carcinoma. Tumor cells similar to basal cells continue to infiltrate and proliferate in the oral mucosal epithelium while forming large and small nests.

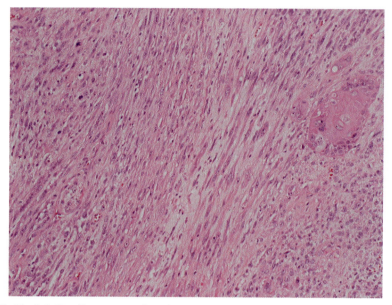

FIG. 14 Spindle cell squamous cell carcinoma. Tumor cells, mainly spindle cells, are interlaced and present with a sarcoma-like appearance. Some cases are associated with normal-type squamous cell carcinoma.

positive for cytokeratin, a marker for epithelial cells, and vimentin, a marker for mesenchymal cells. Macroscopically, it often presents with stalked, broad-based polyps with erosions on the mucosal surface and is most common in the larynx of men in their 70s. In the oral cavity, it occurs in the gums, lips, and tongue. The prognosis is relatively good, and the 5-year survival rate is reported to range from 65% to 95% (Fig. 14).

5.3 Adenosquamous carcinoma

Histologically, adenosquamous carcinoma has the characteristics of both squamous cell cancer and adenocarcinoma. Regular SCC exists on the surface layer, and adenocarcinoma components are often seen in the deep layer. It can occur anywhere in the oral cavity; however, the frequency is low. Adenosquamous carcinoma occurs most often in men in their 60s and 70s and is more malignant than SCC [13, 14]. Regional lymph node metastasis is observed in 75%, and distant metastasis to other sites, such as the lungs, is observed in 25%. The 5-year survival rate is 15%–25%.

5.4 Carcinoma cuniculatum

Carcinoma cuniculatum is a rare subtype of SCC. Histologically, it is characterized by stratified squamous epithelium with few atypical cells.

It proliferates while forming structures, such as densely arranged lumen with keratinized substances or foramen with continuous branching depressions from the surface to a deep part, or while showing papillary growth. Carcinoma cuniculatum is considered to be a well-differentiated, low-grade cancer and is grossly similar to verrucous cancer; nevertheless, the keratinization of the surface layer is not conspicuous. Metastasis is rare; however, local recurrence may occur [15].

5.5 Verrucous SCC

Verrucous SCC is a cancer with markedly keratinized stratified squamous epithelium showing a papillary, verrucous, outward growth [16, 17]. The epithelium extends to a thick epithelial process and exhibits a partially flat basal plane. Tumor tissue is poorly atypical and shows exclusionary growth toward the subepithelium. It is a highly differentiated, low-grade cancer. Clinically, verrucous SCC is broad-based and warty and grows slowly externally. It is rarely accompanied by surface bleeding or ulcer. Most of the tumors occur in the oral cavity, and it is common in older men. The 5-year survival rate is reported to range from 80% to 90% (Fig. 15).

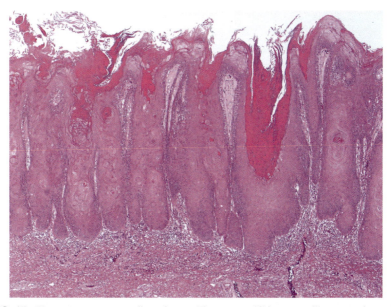

FIG. 15 Verrucous squamous cell carcinoma. The tumor exhibits a wart-like outward growth. Tumor cells are poorly atypical; the surface layer shows thick parakeratosis; and connective tissue is pushed away in the deep part.

5.6 Lymphoepithelial carcinoma

Lymphoepithelial carcinoma is a tumor that involves proliferation of undifferentiated cells with a high nuclear/cytoplasmic ratio and grows with lymphocyte interstitium while forming irregular and undefined tumor nests. Tumor cells exhibit the appearance of poorly differentiated SCC or undifferentiated carcinoma. It has been shown to be associated with EB virus infection, with more than 90% of those that develop in the head and neck region occurring in the tonsils and base of the tongue. Lymph node metastasis is seen from the first visit in approximately 70% of cases. Local control is considered to be high in this tumor with high radiosensitivity [18].

5.7 Papillary SCC

Papillary SCC is a tumor that shows an outward, papillary growth. Histologically, papillary growth of the squamous epithelium is seen with the axis of fibrous connective tissue [19, 20]. The surface layer is mildly keratinized, and the tumor cells show various degrees of cell atypia; however, the infiltration of the tumor cells into the subepithelium is not clear. Papillary SCC occurs most often in the hypopharynx and larynx of men in their 60s and 70s and is rare in the oral cavity. Its prognosis is good (Fig. 16).

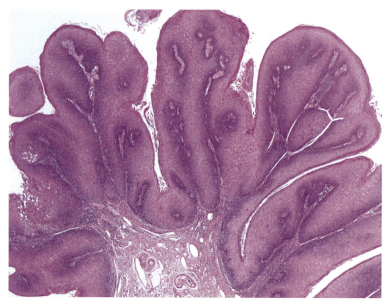

FIG. 16 Papillary squamous cell carcinoma. The tumor grows extroverted and papillary. The keratinization of the surface layer is not remarkable.

Diagnosis of oral squamous cell carcinomas and precancerous lesions Chapter | 2 39

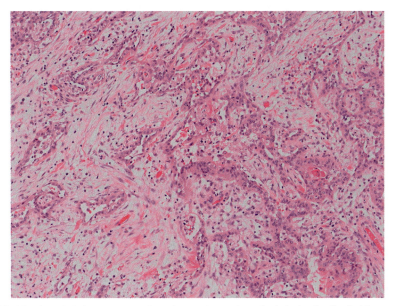

FIG. 17 Acantholytic squamous cell carcinoma. Tumor showed tubular lumina lined with several layers of epithelium. The lumina were filled with acantholytic cells.

5.8 Acantholytic SCC

Acantholytic SCC is a rare subtype of SCC characterized by acantholysis of tumor cells. Histologically, it appears as SCC; however, ductal or vascular-like structures are seen in the nest owing to the removal of the central part of the tumor nest in relation to acantholysis. Tumor cells that have lost intracellular connections can be found in the nests with acantholysis. In the head and neck, it is reported to occur frequently on the lips [21, 22]. Its prognosis is similar to that of SCC (Fig. 17).

6. Conclusion

The 2017 WHO Head and Neck Cancer Classification described OPMDs as a new clinical disease. In addition, epithelial dysplasia in the oral cavity was defined as OED to distinguish it from epithelial dysplasia that occurs at other sites. In this chapter, we explained the concept and classification of OPMDs and OED according to the WHO classification and SCC that comprises the main component of malignant epithelial tumors. We also discussed oral CIS, which is not explained in detail in the WHO classification.

Conflict of interest statement

There are no potential conflicts of interest to disclose.

40 Inflammation and oral cancer

References

[1] Ferlay J, Colombet M, Soerjomataram I, Mathers C, Parkin DM, Piñeros M, et al. Estimating the global cancer incidence and mortality in 2018: GLOBOCAN sources and methods. Int J Cancer 2019;144(8):1941–53.

[2] Warnakulasuriya S. Global epidemiology of oral and oropharyngeal cancer. Oral Oncol 2009;45(4–5):309–16.

[3] Shield KD, Ferlay J, Jemal A, Sankaranarayanan R, Chaturvedi AK, Bray F, et al. The global incidence of lip, oral cavity, and pharyngeal cancers by subsite in 2012. CA Cancer J Clin 2017 Jan;67(1):51–64.

[4] National Center for Chronic Disease Prevention and Health Promotion (US) Office on Smoking and Health. The Health Consequences of Smoking—50 Years of Progress: A Report of the Surgeon General. Atlanta, GA: Centers for Disease Control and Prevention (US); 2014. PMID: 24455788 Bookshelf ID: NBK179276.

[5] Reibel J, Gale N, Hille J, Hunt JL, Lingen M, Muller S, et al. Oral potentially malignant disorders and oral epithelial dysplasia. In: Adel K, et al., editors. WHO classification of head and neck tumors. 4th ed. Lyon: IARC; 2017. p. 112–5.

[6] Sloan P, Gale N, Hunter K, Lingen M, Nylander K, Reibel J, et al. Malignant surface epithelial tumors. In: Adel K, et al., editors. WHO classification of head and neck tumors. 4th ed. Lyon: IARC; 2017. p. 109–11.

[7] Barnes L, Eveson JW, Reichart P, et al, editors. WHO classification of tumours, pathology and genetics of head and neck tumors. 3rd ed. Lyon: IARC Press; 2005.

[8] JSOP Working Committee. Carcinoma in-situ of the oral mucosa: its pathological diagnostic concept based on the recognition of histological varieties proposed in the JSOP oral CIS catalog. J Oral Maxillofac Surg Med Pathol 2014. https://doi.org/10.1016/j.ajoms.2013.11.003.

[9] Fritsch VA, Gerry DR, Lentsch EJ. Basaloid squamous cell carcinoma of the oral cavity: an analysis of 92 cases. Laryngoscope 2014 Jul;124(7):1573–8.

[10] Xie S, Bredell M, Yang H, Shen S, Yang H. Basaloid squamous cell carcinoma of the maxillary gingiva: a case report and review of the literature. Oncol Lett 2014 Sep;8(3):1287–90.

[11] Bice TC, Tran V, Merkley MA, Newlands SD, van der Sloot PG, Wu S, Miller MC. Disease-specific survival with spindle cell carcinoma of the head and neck. Otolaryngol Head Neck Surg 2015 Dec;153(6):973–80.

[12] Gerry D, Fritsch VA, Lentsch EJ. Spindle cell carcinoma of the upper aerodigestive tract: an analysis of 341 cases with comparison to conventional squamous cell carcinoma. Ann Otol Rhinol Laryngol 2014 Aug;123(8):576–83.

[13] Kass JI, Lee SC, Abberbock S, Seethala RR, Duvvuri U. Adenosquamous carcinoma of the head and neck: molecular analysis using CRTC-MAML FISH and survival comparison with paired conventional squamous cell carcinoma. Laryngoscope 2015 Nov;125(11):E371–6.

[14] Masand RP, El-Mofty SK, Ma XJ, Luo Y, Flanagan JJ, Lewis Jr JS. Adenosquamous carcinoma of the head and neck: relationship to human papillomavirus and review of the literature. Head Neck Pathol 2011 Jun;5(2):108–16.

[15] Sun Y, Kuyama K, Burkhardt A, Yamamoto H. Clinicopathological evaluation of carcinoma cuniculatum: a variant of oral squamous cell carcinoma. J Oral Pathol Med 2012 Apr;41(4):303–8.

[16] Mallick S, Breta M, Gupta SD, Dinda AK, Mohanty BK, Singh MK. Angiogenesis, proliferative activity and DNA ploidy in oral verrucous carcinoma: a comparative study including verrucous hyperplasia and squamous cell carcinoma. Pathol Oncol Res 2015 Sep;21(4):1249–57.

Diagnosis of oral squamous cell carcinomas and precancerous lesions **Chapter | 2** **41**

[17] Odar K, Zidar N, Bonin S, Gale N, Cardesa A, Stanta G. Desmosomes in verrucous carcinoma of the head and neck. Histol Histopathol 2012 Apr;27(4):467–74.

[18] Rytkönen AE, Hirvikoski PP, Salo TA. Lymphoepithelial carcinoma: two case reports and a systematic review of oral and sinonasal cases. Head Neck Pathol 2011 Dec;5(4):327–34.

[19] Samman M, Wood HM, Conway C, Stead L, Daly C, Chalkley R, Berri S, Senguven B, Ross L, Egan P, Chengot P, Ong TK, Pentenero M, Gandolfo S, Cassenti A, Cassoni P, Al Ajlan A, Samkari A, Barrett W, MacLennan K, High A, Rabbitts P. A novel genomic signature reclassifies an oral cancer subtype. Int J Cancer 2015 Nov 15;137(10):2364–73.

[20] Ding Y, Ma L, Shi L, Feng J, Liu W, Zhou Z. Papillary squamous cell carcinoma of the oral mucosa: a clinicopathologic and immunohistochemical study of 12 cases and literature review. Ann Diagn Pathol 2013 Feb;17(1):18–21.

[21] Allon I, Abba M, Kaplan I, Livoff A, Zaguri A, Nahlieli O, Vered M. Oral variant of acantholytic squamous cell carcinoma-histochemical and immunohistochemical features. Acta Histochem 2019 Nov;121(8):151443.

[22] Kim JE, Lee C, Oh KY, Huh KH. A rare acantholytic variant of squamous cell carcinoma of the maxilla: a case report and literature review. Medicine (Baltimore) 2020 Aug 7;99(32), e21631.

Chapter 3

Oral cytology in Japan: Its useful approach and criteria for early detection of carcinoma and precursor lesions

Motohiko Nagayama[a], Akinori Ihara[b], and Yoichi Tanaka[c]

[a]*Department of Oral Pathology, Asahi University School of Dentistry, Mizuho, Gifu, Japan,*
[b]*Ihara-Machinaka Dental Clinic, Numazu, Shizuoka, Japan, [c]Department of Surgical Pathology,*
Tokyo Dental College Ichikawa General Hospital, Ichikawa, Chiba, Japan

1. Introduction

Oral squamous cell carcinoma (OSCC) is the most frequent type of oral malignancy and has recently increased in Japan, but its precursor lesions are diverse and often preceded by lesions and conditions collectively referred to as oral potentially malignant diseases (OPMDs) [1]. For appropriative diagnosis and treatment, scalpel biopsy has been the definitive procedure to differentiate these cancers and precancerous conditions, but due to the surgical bio-invasiveness involved, there is often hesitation to use it as a screening test for early detection and treatment. Oral mucosal exfoliative cytology (OMEC) can be used as an adjunctive test for the early detection of oral cancer and OPMDs and for screening decisions for other lesions in a simplified and minimally invasive manner.

In order to understand OMEC, it is important to know the structural and functional characteristics of the oral mucosa that differentiate it from other organs. Histologically, the oral mucosal epithelium is a stratified squamous epithelium with four basic layers: basal, spinous, granular (middle or transparent layer), and superficial keratinized layers, respectively. The thickness of the superficial keratinized layer varies in response to the location and function of the oral mucosa. As keratinization progresses, the granular layer, which represents the layer containing keratohyaline granules, appears, but in parakeratotic and nonkeratinized mucosal epithelia, the granular layer may not appear. Since these oral mucosa are repeatedly exposed to a background of external forces from chewing, prosthetic materials, hot beverages, and other physical and chemical stimuli, OMEC often shows atypical cells as a reactive condition, including

Inflammation and Oral Cancer. https://doi.org/10.1016/B978-0-323-88526-3.00003-8
Copyright © 2022 Elsevier Inc. All rights reserved.

44 Inflammation and oral cancer

fibrin, inflammatory cells, oral bacteria, and ulcerative lesions such as aphtha. OMEC with direct smear method was developed as a screening test for oral mucosa [2,3]. However, there were differences in the cytodiagnosis and handling of OMEC at individual institutions. Several adjunctive aids that improve early detection of OSCC and OPMDs including oral epithelial dysplasia (OED) have been developed; however, there is not yet a proven method with high sensitivity and specificity. Liquid-based cytology (LBC) using appropriative equipment such as a brush has gradually replaced conventional OMEC and represents a well accepted and minimally invasive approach for harvesting cells from the oral mucosa. LBC provides both higher sensitivity and specificity, which shows great potential as a reproducible and reliable tool for the cytodiagnosis and early detection of oral mucosal lesions including OPMDs or OSCCs [4]. On the other hand, the oral mucosa can be observed directly and lesions can be collected directly by biopsy. Specimens can be collected at an early stage when lesions are detected, but the examinations that can be performed in general dental clinics have been limited [5, 6].

2. Procedure of cell collection

Specimens are collected by scraping with an interdental brush or a brush specially designed for the LBC method. After collection, cells can be smeared directly onto a glass slide coated with an antipeeling agent (conventional method) or collected in a special LBC storage solution. In the conventional method, about 80% of the collected cells are lost during the specimen preparation process, but in the LBC method, almost 100% of the cells can be recovered and observed. Since the oral mucosa is frequently affected by erosion and ulceration in response to all kinds of stimuli, and is often accompanied by fibrin deposition, bleeding, or bacterial contamination in the background, the LBC method is recommended to reduce the background conditions that disturb the observation of cellular findings. In some cases, when scraping with a brush is difficult due to easily bleeding lesions, swabs or other instruments may be used, but the amount of cells collected may be reduced. Cells are generally stained with Papanicolaou (Pap) staining or May-Giemsa (MG) staining when salivary gland or hemato-lymphocytic tumors are suspected. If a fungal infection such as *Candida* is suspected, PAS staining can be used for easy detection. Since bacteria are common in the oral cavity, it is often difficult to determine the presence of bacterial colonies and food residues in the background when the oral cavity is poorly cleaned. In such cases, the mouth should be cleaned beforehand with a rinse or, if necessary, a cotton ball or sponge. If the patient complains of pain in the lesion area, rubbing with topical anesthesia such as xylocaine jelly does not affect the cytological image. Since bleeding during collection makes cytodiagnosis difficult, the use of a pretreatment solution is recommended for the LBC method [6].

3. The guideline diagnostic criteria for oral mucosal exfoliative cytology (OMEC)

There are no international/universal diagnostic criteria for OMEC, and they are not the same as the gynecological criteria. The Japanese Society of Clinical Cytology established a diagnostic guideline committee for oral cytology in 2013, and published a Japanese diagnostic guideline for oral cytology in 2015 [7]. In this guideline, the cellular diagnosis and predictive lesion should be described for the evaluation of specimen adequacy, the decision category regarding the presence or absence of atypical cells, and the abnormal cells in the lesion. It is desirable to clearly state the reasons for specimens that are considered inadequate [7].

4. Appropriate evaluation of specimens

In oral mucosa, the degree of keratinization differs depending on the site, such as masticatory mucosa, coated mucosa, and special mucosa, and the number of cells collected from masticatory mucosa is often small due to keratinization. The number of cells collected is often less than 2000 with the conventional method and less than 5000 with the LBC method, and the number of cells collected is often reduced, especially in keratotic lesions. Therefore, the appropriate standard for cervical specimens cannot be directly applied to oral mucosal cytology. However, if the amount of cells collected is extremely small or if there is strong degeneration, the specimen should be reexamined as inappropriate and efforts should be made to increase the amount of cells collected [6, 7].

5. Cytological classification

In the new reporting format by the Working Group on Oral Cytology of the JSCC, it was impossible to adapt the Class classification used in the guideline of cytology of the cervix to the oral cavity. This is because neoplastic changes in the mucosal epithelium of the cervix are mainly caused by squamous metaplasia, and the pathogenesis caused by human papilloma virus (HPV) infection is different from that of the oral mucosa. In addition, many cases are classified as false-positive and do not progress to the next clinical step such as biopsy after cytodiagnosis. Recently, the novel guideline for oral/oropharyngeal cytologic grading system analogous to the Bethesda System for reporting cervical cytology [6]. The new guideline and reporting format in the JSCC 2015 does not recommend the previous Papanicolaou Class classification, but is based on a modification of the Bethesda classification of the cervix. In addition, the JSCC 2015 emphasized (1) evaluation of specimen unsuitability, (2) use of descriptive terms aimed at presumptive diagnosis, and (3) addition of educational notes and suggestions (Table 1, Fig. 1) [7]. It is recommended that the new reporting format should be used in conjunction with [1,6] the new classification:

TABLE 1 Diagnostic criteria for oral exfoliative cytology [7].

Inadequate: insufficient cells number, dried, poorly fixed, destructed specimens
Adequate:
NILM, negative for intraepithelial lesion or malignancy
OLSIL, oral low-grade squamous intraepithelial lesion or low-grade dysplasia
OHSIL, oral high-grade squamous intraepithelial lesion or high-grade dysplasia
SCC, squamous cell carcinoma
IFN, indefinite for neoplasia

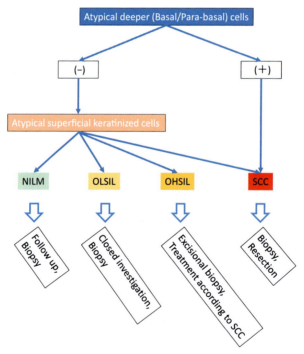

FIG. 1 Cytodiagnostic flowchart and treatment plan. General practitioners/dentists should attempt to screen the patient and follow up with the NILM. In case of further cytological diagnosis, specialists and higher medical institutions for cytology should be responsible. IFN should be handled individually.

1) *NILM (negative for intraepithelial lesion or malignancy)*: Normal or reactive growth with no intraepithelial neoplastic lesion or malignant change. This corresponds to Class I to II in the conventional classification.
2) *OLSIL (oral low-grade squamous intraepithelial lesion of low-grade epithelial dysplasia)*: Oral low-grade intraepithelial tumor is classified as low

atypical epithelial dysplasia or mild to moderate epithelial dysplasia according to the histological classification [1]. In the conventional classification, it corresponds to Class IIb to III.

3) *OHSIL (oral high-grade squamous intraepithelial lesion of low-grade epithelial dysplasia)*: Oral high-grade intraepithelial tumor, corresponding to high atypical epithelial dysplasia/intraepithelial carcinoma in the histological classification [1]. The conventional classification is Class III. In the conventional classification, it corresponds to Class IIIb to IV.

4) *SCC (squamous cell carcinoma)*: Squamous cell carcinoma is indicated by the same notation as in the tissue classification [1]. It corresponds to Class V in the conventional classification.

5) *IFN (indefinite for neoplasia)*: Cytologically, it is difficult to determine the classification of neoplastic or nonneoplastic, such as reactive atypia (Figs. 2–5).

The new classification is characterized by OLSIL and OHSIL, which are intended to correct the inclusion of SCC cells in a small number of Class II cases and to accurately determine neoplastic changes by cytology and to guide the next steps. The OHSIL detects differentiated intraepithelial carcinomas that show keratinization in the superficial layer but various atypia in the deeper layers of the epithelium from keratinized superficial cells, and recommends early aggressive treatment at higher medical institutions more quickly than the conventional Class IV [7].

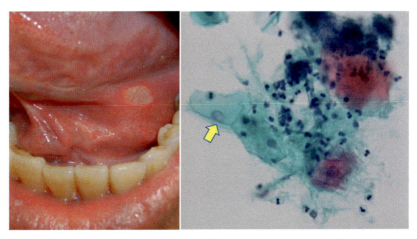

FIG. 2 NILM, 49-year-old man, aphthae. (*Left*) Distinctive ulcer formation with circled redness at the sublingual region; (*Right*): inflammatory background with leukocytes and fibrin precipitation, superficial nonkeratotic epithelial cells have enlarged nuclei but are not hyperchromasia (*arrow*). *LBC*, Pap staining. (*Left*) Reprinted with permission from New oral pathology. 3rd ed. Tokyo: Ishiyaku; 2021.

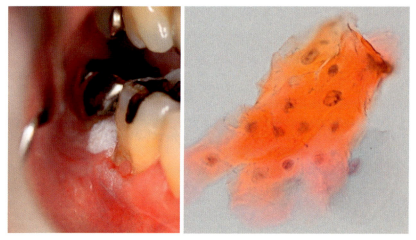

FIG. 3 OLSIL, 59-year-old man. (*Left*) White nonhomogenous/verrucous lesion at the inter dental papillae gingiva; (*Right*) superficial keratotic atypical cells showing high brightness of Orange G cytoplasm, enlargement with increase of chromatin in the nuclei. *LBC*, Pap staining.

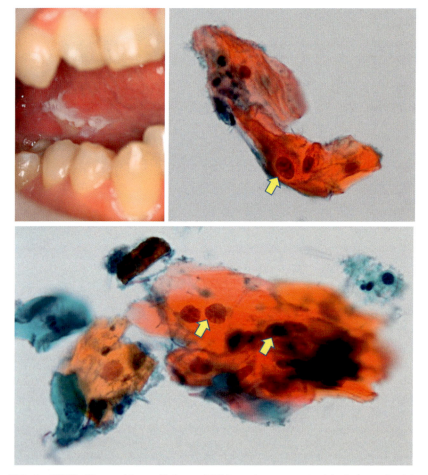

FIG. 4 OHSIL, 56-year-old man. (*Left upper*) White nonhomogenous/nodular lesion at the right tongue edge; (*Right upper and lower*) superficial keratotic atypical cells show dark Orange G cytoplasm and irregular and enlarged nuclei with hyperchromasia (*arrows*). *LBC*, Pap staining.

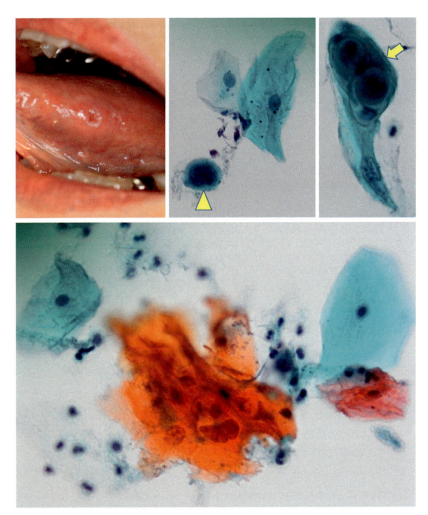

FIG. 5 SCC, 73-year-old woman. (*Left upper*) Nodular lesion with ulcer formation at the left tongue edge; cytological findings are prominent cellular atypia showing cellular diversity; (*Middle upper*) atypical deeper (basal/parabasal) cells show abnormal variation in nuclear shape and size (*arrowhead*); (*Right upper*) the cluster folded together shows "cell in cell/cannibalism" (*arrow*); (*Lower*) superficial keratotic atypical cells show prominent Orange G glistening irregular cytoplasm and enlargement of nuclei with hyperchromasia in the inflammatory background. *LBC*, Pap staining.

5.1 Correspondence between the histological grading system for epithelial dysplasia and the cytodiagnostic criteria

The WHO Classification of Head and Neck Tumors [1] classifies intraepithelial neoplastic lesions of the oral cavity as OED, which are precursor lesions to cancer [1]. OED is defined as "a spectrum of histological and cytological atypia of the epithelium with an increased risk of progression to squamous

50 Inflammation and oral cancer

cell carcinoma, caused by the accumulation of genetic mutations" (Table 2). Depending on the degree of atypia, OED can be classified into three levels: mild dysplasia, moderate dysplasia, and severe dysplasia or in order to be consistent with the clinical context, two stages (binary system) are also proposed: low-grade dysplasia and high-grade dysplasia. The relationship between diagnostic criteria for epithelial dysplasia and cytological criteria is shown in Table 3. This table shows that OLSIL corresponds to mild/moderate/low-grade dysplasia, and OHSIL corresponds to severe/high-grade dysplasia/carcinoma in situ, respectively [7].

TABLE 2 Diagnostic criteria for epithelial dysplasia [1].

Architectural changes	Cytological changes
Irregular epithelial stratification	Abnormal variation in nuclear size
Loss of polarity of basal cells	Abnormal variation in nuclear shape
Drop-shaped rete ridges	Abnormal variation in cell size
Increased number of mitotic figures	Abnormal variation in cell shape
Abnormally superficial mitotic figures	Increased N:C ratio
Premature keratinization in single cells	Atypical mitotic figures
Keratin pearls within rete ridges	Increased number and size of nucleoli
Loss of epithelial cell cohesion	Hyperchromasia

TABLE 3 Grading systems for epithelial dysplasia [1] vs oral exfoliative cytology [7].

WHO dysplasia grade	Binary system	JSCC cytology system
Mild dysplasia	Low-grade dysplasia	OLSIL
Moderate dysplasia		
Severe dysplasia	High-grade dysplasia	OHSIL
	Carcinoma in situ	

5.2 Approach of oral mucosal exfoliative cytology and diagnostic criteria with flow chart [7]

The basic procedures and diagnostic criteria during microscopic observation are shown below:

1) Determine whether the specimen is adequate or inadequate at low magnification.

2) If adequate, observe for the presence of inflammation or other background (clear or cloudy) at low magnification.

3) Examine the type of cells that were collected. Specifically, the quantitative ratio of Orange G-stained keratinizing superficial cells to light-green-stained nonkeratinizing superficial cells and deep-layered cells should be examined from each cell type as follows.

4) If deep-layer cells are present, consider whether the changes are regenerative, degenerative, or neoplastic.

5) If there are no deep-layer cells, carefully observe the superficial cells, either the orange G-stained keratinizing superficial cells or the light-green-stained nonkeratinizing superficial cells. In both types of cells, make a comprehensive judgment by referring to the increase in the degree of cytoplasmic staining (high-luminescence), the nuclear size (swelling/enlargement) and density (amount of chromatin and degree of coarsely), nuclear shape irregularity, nuclear membrane thickening, changes in the number and size of nucleoli, and the overall degree of change (cellular diversity) and background of these changes (Table 4).

TABLE 4 Points for cell observation.

	Superficial cell (Orange G)	Superficial cell (Light-green)	Deep cell (Light-green)
Frequency of occurrence	High/Middle/Low	High/Middle/Low	High/Middle/Low
Abnormal nucleus shape (irregularities)	Presence/Absence	Presence/Absence	Presence/Absence
Amount of nuclear chromatin	More/Less	More/Less	More/Less
Cytoplasmic/ Nuclear diversity (atypia)	Presence/Absence	Presence/Absence	Presence/Absence
Background (contamination)	Clear/Cloudy		
Comprehensive evaluation	Inadequate/NILM/OLSIL/OHSIL/SCC/IFN		

52 Inflammation and oral cancer

5.3 Differential cytodiagnosis

NILM vs OLSIL

OMEC has been used to detects cells of OLSIL neoplastic changes in reactive changes such as hyperkeratosis and epithelial hyperplasia in the clinical state of leukoplakia, which can contribute to early detection. Reactive changes include inflammation associated with aphthous ulcers and infection, regenerative changes associated with ulcers, and changes after radiotherapy and chemotherapy. When a slight increase in N:C ratio, nuclear enlargement, or prominent nucleoli are seen in aggregates, they are often seen as regenerative atypia rather than SCC as seen in single cells. After radiotherapy or chemotherapy in the treatment of cancer, cells with abnormal nuclei and cell enlargement or multinucleated cells are seen in proportion to the treatment. When superficial keratinocytes with prominent Orange G glistening cytoplasm show cellular diversity, including numerous cellular atypia, they should be considered OLSIL, and if they cannot be determined, they should be considered morphologically indistinguishable (IFN) for subsequent short-term follow-up and biopsy.

OHSIL vs SCC

In OHSIL, atypical superficial keratinocytes have more atypical cell aggregates than in OLSIL. The cytoplasm becomes even thicker and the increase in cytoplasmic brightness of the Orange G-favoring cells increases. In SCC, deep layer atypical cells are seen. Atypical keratinocytes often have a spherical shape and show a variety of atypia, including abnormal cell morphology and nuclear atypia [6, 7]. Although cytological diagnosis is used as a screening test to detect oral cancer at an early stage, it is also a feature of this guideline to add educational notes and suggestions after the diagnosis, mainly describing what should be handled by general dentists for early detection and treatment () [7].

5.4 Usefulness, limitations, and future prospects of the LBC method

In the LBC method, cells are smeared onto glass slides using a different principle from that of the conventional method, and it is necessary to be aware of the difference in findings. In cytology using the LBC method, the background is generally clearer, the cells are evenly distributed, and the cells tend to have slightly enlarged cytoplasm and nuclei and slightly fainter nuclear chromatin. However, it should always be kept in mind that there is a frequency of false negatives and false positives in cytology. The quality of DNA extracted from preserved solution (fixer) is better than that of DNA extracted from formalin-fixed paraffin-embedded sections, and DNA can be stored for more than 1 month at room temperature if the cells in the storage solution are fixed in a stable condition. It is also possible to perform next-generation sequencing using RNA from LBC specimens. The LBC method can also be applied to immunostaining

TABLE 5 Adjunctive techniques applied to brush biopsy specimen.

Quantitative cytomorphology

Nuclear DNA content analysis

Immunohistochemical tumor marker identification (immunocytochemistry)

Molecular analysis (RNA extraction and RT-PCR)

Epigenetic alterations (promotor hypermethylation)

Genomic instability and loss of heterozygosity (LOH)

Microsatellite instability (MSI) and restriction fragment length polymorphism (RFLP)

without the special antigen activation treatment, as is the case with paraffin-embedded sections, allowing antibodies to penetrate into cells. Thus, it can be used not only for morphological cell determination but also for molecular biological analysis (Table 5). In the near future, it is expected that artificial intelligence (AI) will be used to make cytodiagnoses based on deep learning that corresponds to the cellular diversity of chromatin quantity and distribution from nuclear and cellular morphology, leading to an increase in cytoplasmic brightness (high-luminescence), so it will be possible to detect atypical cells with high accuracy by automated screening in the near future [6, 8, 9].

References

[1] El-Naggar AK, Chan JKC, Grandis JR, Takata T, Slootweg PJ, editors. WHO classification of head & neck tumours. 4th ed. Lyon: IARC Press; 2017.

[2] Sciubba JJ. Improving detection of precancerous and cancerous oral lesions. Computer-assisted analysis of the oral brush biopsy. U.S. Collaborative OralCDx Study Group. J Am Dent Assoc 1999;130(10):1445–57.

[3] Mehrotra R, Mishra S, Singh M, Singh M. The efficacy of oral brush biopsy with computer-assisted analysis in identifying precancerous and cancerous lesions. Head Neck Oncol 2011 Aug 24;3:39.

[4] Navone R, Burlo P, Pich A, Pentenero M, Broccoletti R, Marsico A, Gandolfo S. The impact of liquid-based oral cytology on the diagnosis of oral squamous dysplasia and carcinoma. Cytopathology 2007 Dec;18(6):356–60.

[5] Alsarraf A, Kujan O, Farah CS. Liquid-based oral brush cytology in the diagnosis of oral leukoplakia using a modified Bethesda cytology system. J Oral Pathol Med 2018 Oct;47(9):887–94.

[6] Mehtotra R, editor. Oral cytology a concise guide. New York, USA: Springer; 2013.

[7] Yoichi T, Japanese Society for Clinical Cytology, et al. Oral. In: Japanese Society for Clinical Cytology, editor. JSCC Atlas and Guidelines for Cytopathologycal Diagnosis 5, Digestive organs. 1st ed. Tokyo, Japan: Kanehara; 2015. p. 18–79.

[8] Reynolds JP, Zhou Y, Jakubowski MA, Wang Z, Brainard JA, Klein RD, Farver CF, Almeida FA, Cheng YW. Next-generation sequencing of liquid-based cytology non-small cell lung cancer samples. Cancer Cytopathol 2017 Mar;125(3):178–87.

54 Inflammation and oral cancer

[9] Sunny S, Baby A, James B, Balaji D, Aparna V, Rana M, Gurpur P, Skandarajah A, D'Ambrosio M, Ramanajinappa R, Pm S, Raghavan N, Kandasarma U, Sangeetha N, Raghavan S, Hedne N, Koch F, Fletcher D, Selvam S, Kuriakose M. A smart tele-cytology point-of-care platform for oral cancer screening. PLoS One 2019;14(11), e0224885.

Chapter 4

Imaging findings of oral cancers

Hiroki Kato and Masayuki Matsuo
Department of Radiology, Gifu University Graduate School of Medicine, Gifu, Japan

1. Introduction

Squamous cell carcinomas (SCCs) account for approximately 90% of the malignancies arising from various subsites of the oral cavity. Other malignancies in this region can arise from the epithelium, lymphoid tissue (lymphoma), connective tissue (sarcoma), minor salivary glands (salivary gland cancer), melanocytes (mucosal melanoma), and other sites (metastasis from a distant tumor).

Oral cancer undoubtedly has a multifaceted etiology. The major risk factors for oral cancer include tobacco smoking; alcohol abuse; human papillomavirus infection; Epstein–Barr virus infection; candidiasis; precancerous lesions; industrial pollution; occupational hazards; inadequate oral hygiene; improper dental prostheses; diet low in vitamins, fruits, and vegetables; UV-radiation exposure; pale skin complexion; immunological defects; and Plummer-Vinson syndrome [1].

Oral cancers are treated with surgery, chemotherapy, and radiotherapy depending on the histological subtype, location, and clinical stage at diagnosis. Therefore, accurate staging of oral cancer based on sufficient knowledge of the anatomy and the most common routes of spread is essential to allow for optimal treatment planning. Inaccurate assessment of the tumor extension or clinical stage may lead to unnecessary or incomplete surgical interventions or radiotherapy, resulting in a negative outcome [2].

Imaging of the oral cavity can be limited by oral mucosal surface lesions and metallic artifacts from dental implants; therefore, oral cancer may be screened on clinical examination. However, imaging modalities and protocols can be tailored according to an individual patient's specific presentation using a combination of ultrasonography, CT, and MR imaging [3]. Consequently, radiological imaging plays a critical role in staging of oral cancer, determining the deep margins for either tumor resectability or radiation planning, status of the jaw, teeth, and regional lymph nodes, and assessing secondary primary tumors [4]. Thus, this chapter describes the fundamental roles of imaging in managing oral cancer, the anatomy of the oral cavity, the 8th edition American Joint Committee

Inflammation and Oral Cancer. https://doi.org/10.1016/B978-0-323-88526-3.00004-X
Copyright © 2022 Elsevier Inc. All rights reserved.

56 Inflammation and oral cancer

on Cancer (AJCC) and Union for International Cancer Control (UICC) staging system, the imaging modalities and protocols, the imaging findings at each subsite, and the patterns of spread.

2. Anatomy of the oral cavity

The oral cavity is separated from the oropharynx by an imaginary line drawn across the circumvallate papillae, the tonsillar pillars, and the junction of the hard and soft palates. The remaining borders of the oral cavity include the lips, anteriorly; the cheeks, laterally; the hard palate, superior alveolar ridge, and teeth, superiorly; and the inferior alveolar ridge, teeth, and the mylohyoid muscle, inferiorly [4]. The anatomical subsites of oral cavity are the mucosal lips, anterior two-thirds of the tongue (oral tongue), buccal mucosa, upper and lower alveolar ridge, floor of the mouth, hard palate, and retromolar gingiva (retromolar trigone). Therefore, oral cancers are usually classified into the following seven subsites: (1) lip, (2) oral tongue, (3) buccal mucosa, (4) upper and lower gingiva, (5) floor of the mouth, (6) hard palate, and (7) retromolar trigone.

3. Staging of oral cancer

Oral cancers are classified according to the tumor, node, metastasis (TNM) criteria of the AJCC and UICC to aid in establishing a modern and appropriate therapy. Conventionally, T1–T3 categories are purely based on the greatest tumor dimension (T1, ≤ 2 cm; T2, > 2 cm to ≤ 4 cm; and T3, > 4 cm). However, the T categorization of oral cancer in the 8th edition of the AJCC and UICC criteria was modified by incorporating the depth of invasion (DOI) because DOI is the most important negative prognostic factor and strongly associated with regional nodal metastasis. The National Comprehensive Cancer Network (NCCN) recommended neck dissection in patients with DOI greater than 4 mm. Each 5 mm increase in DOI raises the tumor by one T category, and DOI is classified into superficial and less invasive (≤ 5 mm), moderately invasive (> 5 mm to ≤ 10 mm), and deeply invasive (> 10 mm) for clinical staging purposes (Table 1). T4a tumors (oral cavity) are moderately advanced local diseases and involve the bone, maxillary sinus, or skin of the face. Superficial erosion of bone/tooth socket alone by a primary gingival cancer is not sufficient to classify a tumor as T4. T4b tumors are very advanced local diseases and involve the masticator space, pterygoid plates, and skull base, or encase the internal carotid artery.

The TNM classification accounts for the metastatic burden in lymph nodes by stratifying the N stage using the size, number, and laterality of involved nodes. Extranodal extension (ENE) of nodal metastasis is a significant prognostic factor of head and neck SCC; therefore, ENE is classified as N3b by the 8th edition of AJCC and UICC criteria (Table 2). The N categorization is determined by pathological and clinical ENE; however, to be able to assist with treatment planning in the clinical setting, radiologists need to be able to provide more information, such as the extent of the ENE and its resectability [5].

Imaging findings of oral cancers **Chapter | 4   57**

TABLE 1 T categorization of oral cancer in the 8th edition of AJCC and UICC.

T1	Tumor ≤2 cm, DOI ≤5 mm
T2	Tumor ≤2 cm, DOI >5 mm, and ≤10 mm *or* tumor >2 cm but ≤4 cm, DOI ≤10 mm
T3	Tumor >4 cm, DOI ≤10 mm *or* any size, DOI >10 mm
T4a	Moderately advanced local disease (lip) Tumor invades through cortical bone or involves the inferior alveolar nerve, floor of the mouth, or skin of the face (i.e., chin or nose) (oral cavity) Tumor invades adjacent structures only (e.g., through cortical bone of the mandible/maxilla, or involves the maxillary sinus or skin of the face)
T4b	Very advanced local disease Tumor invades masticator space, pterygoid plates, or skull base, and/or encases internal carotid artery

Note: AJCC=American Joint Committee on Cancer, UICC=Union for International Cancer Control, DOI=depth of invasion.

TABLE 2 Clinical N categorization of oral cancer in the 8th edition of AJCC and UICC.

N0		No regional lymph node metastasis
N1		Metastasis in a single ipsilateral lymph node, 3 cm or smaller in greatest dimension without ENE
T2	T2a	Metastasis in a single ipsilateral lymph node, larger than 3 cm but not larger than 6 cm in greatest dimension without ENE
	T2b	Metastasis in multiple ipsilateral lymph nodes, none larger than 6 cm in greatest dimension without ENE
	T2c	Metastasis in bilateral or contralateral lymph nodes, none larger than 6 cm in greatest dimension without ENE
T3	T3a	Metastasis in a lymph node, larger than 6 cm in greatest dimension without ENE
	T3b	Metastasis in any lymph node(s) with clinically overt ENE

Note: AJCC=American Joint Committee on Cancer, UICC=Union for International Cancer Control, ENE=extranodal extension.

58 Inflammation and oral cancer

TABLE 3 M categorization of oral cancer in the 8th edition of AJCC and UICC.

M0	No distant metastasis
M1	Distant metastasis

Note: AJCC = American Joint Committee on Cancer, UICC = Union for International Cancer Control.

Lastly, the M categorization is determined by the presence or absence of distant metastasis, generally in locations to which the cancer spreads by vascular channels or lymphatics (Table 3).

4. Imaging modalities and protocols in oral cancer

Oral cancers are primarily evaluated by clinical examination, and their diagnosis is made by histological examination by biopsy. Cross-sectional imaging is used for staging of oral cancers and allows the visualization of the pathology beneath the mucosa, helps determine the size and depth of the primary tumor, detects invasion of neighboring structures, assesses lymph node metastases, and excludes secondary primary tumors [6]. The specific imaging issues to be addressed include bone involvement and the degree of submucosal extension, which are important for developing treatment strategies and follow-up plans during and after treatment.

4.1 CT

CT (computed tomography) is widely available, readily accessible, and relatively inexpensive compared with MR imaging; therefore, in oral cancers, CT usually serves as a first-line imaging modality used for detecting primary tumors, distinguishing the histological subtypes, and determining the clinical stage [3]. CT is adequate for delineating the size and extent of primary tumors and assessing regional lymph node metastases. Early cortical bone erosion is not well demonstrated on MR images, whereas CT images can provide detailed information about cortical bone involvement; thus, CT should be indicated when bone involvement is clinically suspected [7]. Contrast-enhanced CT can accurately determine regional nodal metastases, which may initially appear as normal lymph nodes [1]. However, despite advances in imaging technology, no imaging methods have reached 100% accuracy, as they cannot detect micrometastases within the lymph nodes of clinically N0 necks [8]. The disadvantages include radiation exposure, the need to inject iodinated contrast medium, poor soft-tissue contrast, and metallic artifacts due to dental fillings (Fig. 1) [6].

Contrast-enhanced CT is performed from the skull base to the upper border of the manubrium sterni. Axial, coronal, and sagittal multiplanar reconstructions are obtained at a section thickness of less than 3 mm. These CT images are reconstructed using both soft-tissue and bone algorithms.

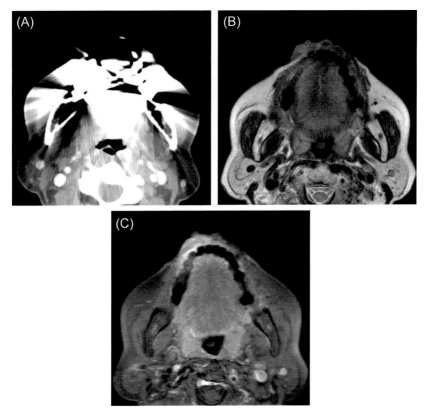

FIG. 1 A 76-year-old woman with squamous cell carcinoma of the right lip (T2). (A) Contrast-enhanced CT image cannot be evaluated due to the severe metallic artifacts caused by dental fillings. (B) T2-weighted image shows an elevated isointense lesion (*arrow*) of the right lip. (C) Fat-suppressed gadolinium-enhanced T1-weighted image shows a heterogeneously enhanced lesion (*arrow*).

4.2 MRI

MR (magnetic resonance) imaging is complementary to CT but offers distinct advantages, including superior soft-tissue contrast, multiplanar capability, and reduced artifacts from dental fillings [6]. MR imaging is superior to CT in detecting and characterizing primary tumors even for exceedingly small lesions, particularly in the oral tongue and floor of the mouth, and in assessing local tumor extent, including early bone marrow involvement, pterygopalatine fossa infiltration, prevertebral muscle involvement, and perineural tumor spread. Therefore, MR images assist in surgical planning, such as choosing the method and scope of surgical resection, and evaluation of complications that may occur during and after surgery. Tumor invasion to the floor of the mouth is particularly well observed on coronal and sagittal MR images, which may provide information on the tongue base involvement and the extent of pharyngeal infiltration

60 Inflammation and oral cancer

[9]. MR imaging entails no radiation, and no iodinated contrast medium must be administered [6]. However, tongue movement with unstable swallowing may cause serious artifacts on MR imaging.

Axial and coronal T1-weighted, T2-weighted, and short TI inversion recovery (STIR) sequences with 3 mm thickness and a field of view of 18–20 cm are standard protocols for oral cancers. Diffusion-weighted (DW) imaging using a b-value of 0 and $1000 \, s \, mm^{-2}$ is preferred because it can precisely detect cancer-involved structures and differentiate between reactive and metastatic lymph nodes. Moreover, DW imaging can help assess a patient's early response to chemoradiotherapy [1]. Furthermore, a fat-suppressed gadolinium-enhanced T1-weighted sequence is performed because it can successfully display tumor extension and perineural spread. Bone marrow involvement is assessed on T1-weighted, STIR, and fat-suppressed contrast-enhanced T1-weighted sequences, whereas regional lymph node metastases are evaluated on T2-weighted, STIR, fat-suppressed contrast-enhanced T1-weighted, and DW sequences with apparent diffusion coefficient maps [10].

4.3 ^{18}F-fluorodeoxyglucose positron emission tomography (PET)/CT

Despite the high precision of ^{18}F-fluorodeoxyglucose PET/CT in detecting primary tumors, it has not replaced CT and MR imaging, particularly in detecting small lesions [11]. PET/CT is a useful tool for assessing both regional nodal and distal metastases and for detecting possible secondary primary tumors [4]. Although PET/CT is clinically used for detecting regional lymph node metastases, existing data on the greater accuracy of PET/CT are controversial, and most authors do not justify the use of PET/CT in patients with head and neck cancer, even in individuals with clinical stage N0 disease [11]. Regarding the differentiation between surgical scars and tumor recurrence, PET/CT is superior to CT and MR imaging. However, because PET/CT has some limitations due to physiological uptake in lymphoid tissue, nasal mucosa, salivary glands, and muscle activity, and in irradiated and postoperative tissues, the use of PET/CT in evaluating recurrent oral cancers warrants careful consideration [4].

5. Imaging findings of oral cancer at each subsite

5.1 Lip

The lip is the most common subsite of oral cancers, accounting for approximately 40% of all cases. Cancer of the lip usually arises from the vermilion border and can spread by lateral extension to the adjacent skin or by deep extension to underlying muscles of the mouth, such as the orbicularis oris. Because lip lesions are easily detected and assessed by direct visualization, radiological imaging plays a role in detecting more advanced lesions with indeterminate margins [2, 4, 6]. The role of CT and MR imaging is in determining tumor extension, including the integrity of the adjacent bone, perineural spread, and regional lymph node metastasis. On CT and MR imaging, a primary tumor may appear as a mass with or without areas of ulceration, and subtle bone erosion usually occurs along the buccal surface of the mandibular or maxillary alveolar ridge (Fig. 1) [6]. Furthermore, CT may be superior for assessing subtle bony erosive changes, whereas MR imaging is superior for assessing bone marrow involvement [4]. However, care must be taken in evaluating MR images because overestimation of the extent of bone marrow involvement may occur due to false-positive findings secondary to hemorrhage and inflammation [4]. If the adjacent bone invasion is observed, the lip cancer becomes upstaged to a T4a, which is a contraindication for surgery [4]. Osseous infiltration causes perineural invasion along the alveolar nerves, making local-regional control particularly difficult [2]. Lymphatic fluid is drained predominantly to the level I and II lymph nodes.

5.2 Oral tongue

MR imaging can clearly show the tongue muscle (intrinsic and extrinsic muscles) and interdigitated fat. The intrinsic muscles of the tongue include the inferior and superior longitudinal, transverse, and vertical muscles. The extrinsic muscles (and their sites of attachment) of the tongue include the genioglossus (mandible), styloglossus (styloid process), palatoglossus (soft palate), and hyoglossus (hyoid bone) muscles. The styloglossus muscle extends from the styloid to the parapharyngeal fat and can be identified as it extends into the posterior tongue and subsequently interweaves into the hyoglossus muscle [4]. Therefore, the adjacent tissues around the styloglossus are possible pathways for tumor spread from the tongue to the skull base [4]. In addition, inferior tumoral extension allows tumor access to the mylohyoid and geniohyoid muscles [4].

SCCs of the oral tongue most commonly involve the middle third and the posterior third of the tongue, with most cases occurring on the lateral surface and undersurface of the tongue. Sometimes, tongue cancers are difficult to identify on CT images, because the tongue is obscured by dental streak artifacts, which are often adequately serious to render imaging studies uninterpretable [7]. Meanwhile, MR imaging is the preferred imaging modality for evaluating tongue cancers because it provides valuable information both within and around the tongue (Fig. 2) [7]. In addition to axial MR images, coronal images may provide additional information about the exact extent of tongue cancers and their relationship to the neurovascular bundle [6].

The DOI is a valuable parameter for predicting regional nodal metastasis and survival in oral cancers. Although radiological DOI is strongly correlated with pathological DOI, it is often 2–3 mm larger than pathological DOI [12]. The cause of this discrepancy may be due to the adjacent inflammation, edema, and fibrosis observed on MR images and due to the shrinking of specimens during formalin fixation [12]. On MR images, the radiological DOI is measured from the horizontal reference line, determined as the line connecting the junctions of the tumor and the normal mucosa at contralateral margins to the deepest aspect of the tumor lesion [12].

62 Inflammation and oral cancer

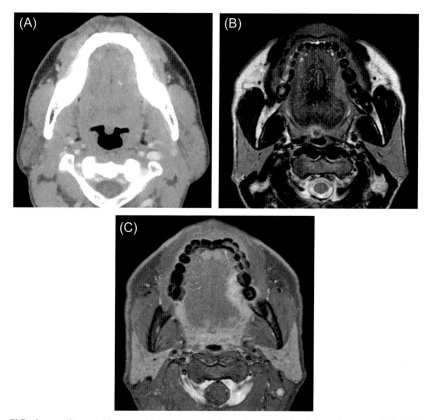

FIG. 2 A 42-year-old man with squamous cell carcinoma of the left oral tongue (T2). (A) Contrast-enhanced CT image shows a slightly enhanced lesion (*arrow*) on the lateral surface of the left tongue. (B) T2-weighted image shows a mild hyperintense lesion (*arrow*) of the left tongue. Radiological depth of invasion on T2-weighted image is measured as 9 mm. (C) Fat-suppressed gadolinium-enhanced T1-weighted image shows a heterogeneously enhanced lesion (*arrow*). The size and depth of invasion are slightly larger than those on T2-weighted image due to the peritumoral inflammation.

Tongue cancers tend to invade the tongue musculature, providing an easy route of spread deep into the intrinsic muscles and along the extrinsic muscles [2]. Among the extrinsic muscles of the tongue, laterally located muscles (the hyoglossus, styloglossus, and palatoglossus muscles) are the first target of invasion by tongue cancers. The involvement of any extrinsic muscle necessitates an evaluation of the muscle at the site of its attachment [2]. In addition, because the neurovascular bundle is particularly vulnerable to invasion by cancers and easily permits perineural extension, the evaluation of the neurovascular bundle is essential. The neurovascular bundle of the tongue comprises the lingual artery (the third branch of the external carotid artery) and veins, the hypoglossal nerve (cranial nerve XII), and the lingual nerve (a branch of the mandibular nerve, which is cranial nerve V) [13]. Anatomically, surgically, and radiologically, the hyoglossus muscle is the most important landmark for locating these neurovascular structures. The lingual artery runs medial to the hyoglossus muscle, whereas the hypoglossal nerve, accompanied by the lingual veins, lies lateral to this muscle [13]; then, they reach successively the lingual pedicle, genioglossus muscle, lingual septum, and contralateral half of the tongue [6]. Evaluation of the midline of the tongue and contralateral neurovascular bundle is important for surgical planning because tumor extension across the midline and involvement of the contralateral neurovascular bundle require total glossectomy or nonsurgical alternatives (combined radiation therapy and chemotherapy) (Fig. 3).

Although the pattern of lymphatic spread of tongue cancers predominantly involves level I and II lymph nodes, "skip metastasis" or "nodal skip metastasis" (e.g., regional spread to level III and IV nodes, without the involvement of level I and II nodes) can rarely occur.

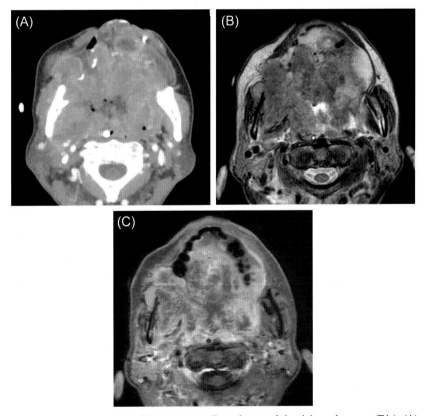

FIG. 3 A 60-year-old man with squamous cell carcinoma of the right oral tongue (T4a). (A) Contrast-enhanced CT image shows a bulky, heterogeneously enhanced lesion (arrows), which crosses the midline of the tongue and occupies the oral cavity. (B) T2-weighted image shows a bulky, heterogeneously iso- to hyperintense lesion (*arrows*) with extension of the right masticator space (*arrowhead*). (C) Fat-suppressed gadolinium-enhanced T1-weighted image shows extensive unenhanced areas suggestive of necrosis (*arrows*) within the tumor.

5.3 Buccal mucosa and gingiva

The buccal mucosa are the mucous membranes lining the inner surface of the lips and cheeks. Buccal cancers commonly occur on the lateral walls of the buccal cavity. The superficial spread of buccal cancers is best assessed clinically; however, the common routes of spread include lateral submucosal extension along the buccinator muscle toward the pterygomandibular raphe and subsequent bony involvement [4].

The gingiva consists of mucosal tissue covering the alveolar processes of the maxilla and mandible inside the mouth. Gingival cancers frequently occur in the molar and premolar regions along the gingival margin of a tooth. The buccal space is the most common site of spread of upper and lower gingival cancers, and the masticator space is the second most common site of spread of cancer arising from the lower gingiva in the molar region [14]. Gingival cancers are associated with advanced stages at the time of diagnosis, due to an early invasion of the contiguous maxillary or mandibular bone.

Evaluation of buccal and gingival cancers should address the extent of submucosal spread, osseous involvement, involvement of the retromolar trigone and pterygomandibular raphe, and cervical lymphatic spread [2]. Tumor involvement of the maxilla may allow spread into the paranasal sinuses (Fig. 4), whereas that of the mandible introduces the possibility of perineural or intramedullary extension (Fig. 5). CT is useful for detecting slight cortical invasion caused by gingival cancers. Although bone marrow invasion is usually assessed using MR imaging, the specificity of MR imaging is significantly lower than that of CT [15]. Gingival cancers may spread into the retromolar trigone and spread along the pterygomandibular raphe, a thick fascial band extending between the posterior border of the mandibular mylohyoid ridge and the hamulus of the medial pterygoid plate. Because the pterygomandibular raphe serves as a

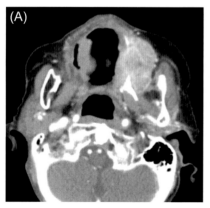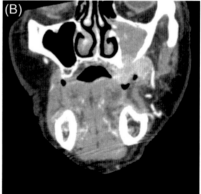

FIG. 4 An 86-year-old woman with squamous cell carcinoma of the left upper gingiva (T4a). (A) Contrast-enhanced CT image shows a heterogeneously enhanced lesion (*arrow*) of the left upper gingiva. (B) Coronal contrast-enhanced CT image shows the tumor extension into the left maxillary sinus (*arrow*).

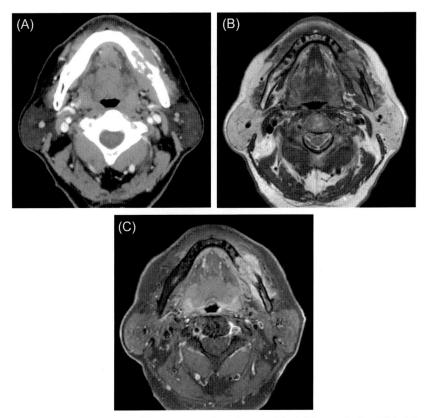

FIG. 5 A 69-year-old man with squamous cell carcinoma of the left lower gingiva (T4a). (A) Contrast-enhanced CT image shows an enhanced lesion (*arrow*) of the left lower gingiva with extensive destruction of the left mandible. (B) T2-weighted image shows a mild hyperintense lesion (*arrow*) with involvement of the left mandible. (C) Fat-suppressed gadolinium-enhanced T1-weighted image shows an enhanced lesion (*arrow*) with involvement of the left mandible.

junction between the oral cavity, oropharynx, and nasopharynx, invasion of the pterygomandibular raphe by oral cancers allows spread into the oropharynx and nasopharynx. Lymphatic spread tends to occur predominantly to level I and II lymph nodes.

5.4 Floor of the mouth

The floor of the mouth is bordered by the lingual aspect of the lower gingiva, anteriorly; the alveolar ridge of the mandible, laterally; the insertion of the anterior tonsillar pillar into the tongue, posteriorly; and the free inferior aspect of the tongue, medially [2]. The sublingual and submandibular glands drain into the floor of the mouth via the major sublingual (Bartholin) ducts and submandibular (Wharton) ducts. The main supporting structure of the floor of the mouth is the mylohyoid muscle, which originates from the mylohyoid line on the internal surface of the mandible and inserts into the median raphe, while the posterior fibers attach to the front of the hyoid bone. The mylohyoid muscle is a key anatomical landmark separating the sublingual space from the submandibular space, although the communication between the sublingual and submandibular spaces is observed by the normal discontinuity within the mylohyoid muscle or along the posterior aspect of the mylohyoid muscle.

Most cancers of the floor of the mouth arise within 2 cm of the anterior midline. In cancers of the floor of the mouth, the involvement of the hyoglossus muscles, mylohyoid muscle, submandibular space, and neurovascular pedicle, mandibular osseous involvement, and cervical lymphatic involvement should be evaluated [2]. Superior spread may involve the ventral surface of the oral tongue; anterior and lateral spread may involve the adjacent gingival mucosa, destroy the lingual cortex, and then involve the marrow of the mandible; inferior spread may infiltrate the genioglossus or mylohyoid muscles; posterior spread often involves the tongue base [7]. Spread beneath the mucosa into the sublingual gland can obstruct the Wharton duct, resulting in submandibular gland sialadenitis, which is often the first clinical manifestation; therefore, careful examination of the floor of the mouth is required in patients with an enlarged submandibular gland [6]. Whether the tumor has invaded through the midline lingual septum or involved the contralateral neurovascular bundles is a determining factor between a hemiglossectomy or total glossectomy [4]. The choice of imaging modality depends on clinical assessment. In the floor of the mouth, MR imaging is superior to CT, regarding the assessment of tumor extension, because of its excellent soft-tissue contrast (Fig. 6). If the primary objective is to demonstrate or rule out mandibular cortical erosion, CT with bone algorithms is recommended.

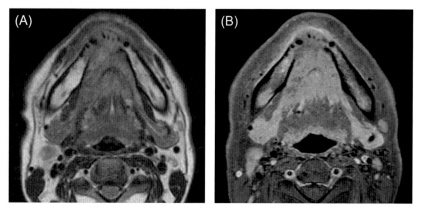

FIG. 6 A 46-year-old man with squamous cell carcinoma of the right floor of the mouth (T4a). (A) T2-weighted image shows a mild hyperintense lesion (*arrow*) of the floor of the mouth with involvement of the mandible (*arrowhead*). (B) Fat-suppressed gadolinium-enhanced T1-weighted image shows a homogeneously enhanced lesion (*arrow*) with involvement of the mandible (*arrowhead*).

5.5 Hard palate

The hard palate is formed by the palatine process of the maxilla and the horizontal plate of the palatine bone covered by a mucous membrane, and hundreds of minor salivary glands are located between the mucosal surface and underlying bone [16]. Palatal tumors commonly arise from the minor salivary glands, and minor salivary gland tumors have an affinity for the posterior hard and soft palates [16]. The most common malignant salivary gland tumor of the palate is adenoid cystic carcinoma, followed by mucoepidermoid carcinoma, adenocarcinoma, and polymorphous low-grade adenocarcinoma [16]. Alternatively, epithelial tumors frequently arise from the soft palate; therefore, primary SCC of the hard palate is rare and often represents extension from an adjacent gingival lesion (Fig. 7).

Hard palate cancers should be assessed for determining the degree of bone invasion and the extent of perineural spread. Advanced tumors may invade the maxilla, nasal cavity, buccal mucosa, tongue, or retromolar trigone. As the mucosa of the hard palate is closely applied to the underlying bone, early osseous erosion is often encountered [7]. Coronal and sagittal CT images with bone algorithms are extremely useful for adequate assessment of adjacent bone involvement because of the axial orientation of the hard palate. Adenoid cystic carcinoma of the palate has a propensity for perineural tumoral spread along the greater and lesser palatine nerves, providing a pathway upward to the pterygopalatine fossa (Fig. 8) [4]. From the pterygopalatine fossa, tumors can travel along the maxillary nerve into the foramen rotundum, anteriorly into the infraorbital foramen, or along the Vidian nerve in the Vidian canal [6].

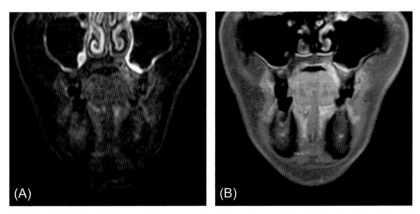

FIG. 7 A 63-year-old woman with squamous cell carcinoma of the left hard palate (T2). (A) Coronal fat-suppressed T2-weighted image shows a mild hyperintense lesion (*arrow*) of the left hard palate. (B) Coronal fat-suppressed gadolinium-enhanced T1-weighted image shows a homogeneously enhanced lesion (*arrow*) without involvement of the left maxilla.

68 Inflammation and oral cancer

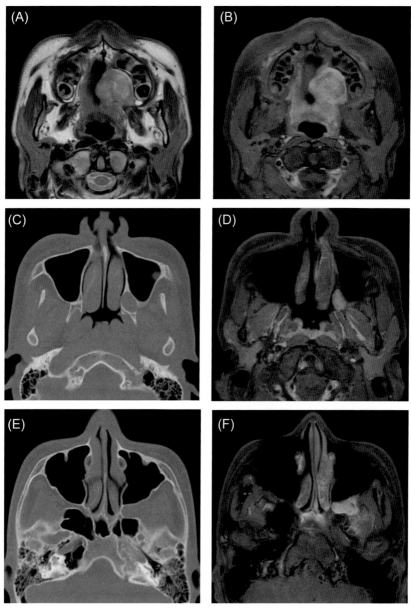

FIG. 8 A 14-year-old man with adenoid cystic carcinoma of the left hard palate (T4a). (A) T2-weighted image shows a heterogeneously hyperintense lesion (*arrow*) of the left hard palate. (B) Fat-suppressed gadolinium-enhanced T1-weighted image shows a homogeneously enhanced lesion (*arrow*) of the left hard palate. (C) Unenhanced CT image with bone algorithm shows an enlarged left greater palatine foramen (*arrow*). (D) Fat-suppressed gadolinium-enhanced T1-weighted image shows a homogeneously enhanced lesion (*arrow*) with extension into the left greater palatine foramen. (E) Unenhanced CT image with bone algorithm shows an enlarged left pterygopalatine fossa (*arrow*). (F) Fat-suppressed gadolinium-enhanced T1-weighted image shows a homogeneously enhanced lesion (*arrow*) with extension into the left pterygopalatine fossa.

Imaging findings of oral cancers Chapter | 4 **69**

Fat-suppressed contrast-enhanced T1-weighted imaging is the best imaging modality for evaluating perineural spread. Lymphatic spread usually occurs along the facial and retropharyngeal lymph nodes and along the upper jugular chain [6].

5.6 Retromolar trigone

The retromolar trigone is a triangular mucosal area situated behind the posterior molars covering the anterior surface of the lower ascending ramus of the mandible. This unique location, which is a junction point between the oral cavity, oropharynx, and nasopharynx, allows for complex tumor spread [4]. Multiple muscles, including the buccinator, orbicularis oris, and superior constrictor muscles insert into the retromolar trigone, allowing different routes for tumor growth [4]. Beneath the mucosal surface of the retromolar trigone is the pterygomandibular raphe, attaching superiorly to the hook of the hamulus arising from the medial pterygoid plate and inferiorly to the posterior mylohyoid line of the mandible [4]. This raphe allows tumor access to the oropharynx, nasopharynx, and skull base.

Retromolar trigone tumors account for 7% of all tumors affecting the oral cavity. Imaging evaluation of the retromolar trigone is critical because tumor extension to this subsite cannot be determined clinically. Evaluation of retromolar trigone cancers should include assessments of submucosal spread, including the involvement of the muscles of mastication, osseous involvement, neurovascular extension, and cervical lymphatic spread [2]. The clinical significance of the retromolar trigone is its ability to provide easy access to numerous routes of spread; thus, cancers originating from the retromolar trigone can spread to any adjacent structure. The tumors often show posterior spread with the involvement of the mandibular ramus, masticator space, superior constrictor muscles, and mandibular branch of the trigeminal nerve (Fig. 9). Anterior spread occurs along the alveolar ridge, and inferior spread occurs along the mandible and inferior alveolar nerve [2]. Furthermore, tumors may spread along the pterygomandibular raphe, providing access superiorly to the temporalis muscle, medially to the pterygomandibular space where the lingual and inferior alveolar nerves run, inferiorly into the floor of the mouth, or anteriorly from the medial pterygoid plate into the pterygopalatine fossa (Fig. 10) [6]. Contrast-enhanced CT allows the evaluation of both soft tissues and bones; however, artifacts due to metallic dental fillings reduce the quality of CT images and increase the difficulty of interpretation. MR imaging is often useful for assessing submucosal spread and regional extent of invasive retromolar trigone tumors because it is less affected by dental filling artifacts.

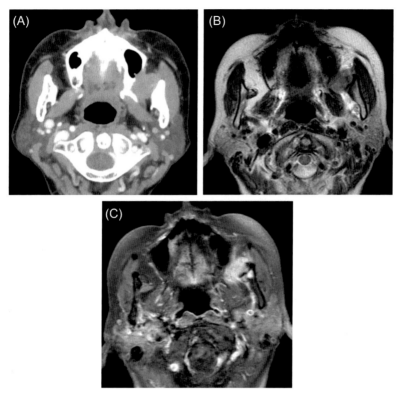

FIG. 9 A 79-year-old woman with squamous cell carcinoma of the left retromolar trigone (T4a). (A) Contrast-enhanced CT image shows an enhanced lesion (*arrow*) of the left retromolar trigone with destruction of the left maxillary sinus wall. (B) T2-weighted image shows a mild hyperintense lesion (*arrow*) of the left retromolar trigone. (C) Fat-suppressed gadolinium-enhanced T1-weighted image shows an enhanced lesion (*arrow*) with involvement of the left mandibular ramus (*arrowhead*).

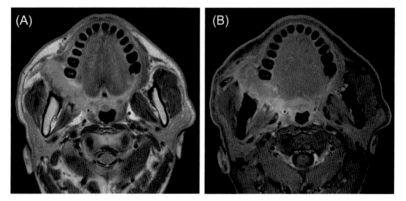

FIG. 10 A 71-year-old man with squamous cell carcinoma of the right retromolar trigone (T4a). (A) T2-weighted image shows a mild hyperintense lesion (*arrow*) of the right retromolar trigone. (B) Fat-suppressed gadolinium-enhanced T1-weighted image shows a mildly enhanced lesion (*arrow*) with extension into the right pterygomandibular raphe (*arrowhead*).

6. Patterns of spread of oral cancer

6.1 Bone invasion

An important role of imaging in evaluating patients with oral cancer is to evaluate for the presence of mandibular invasion [17]. Meanwhile, maxillary invasion is most often observed in cancers of the retromolar trigone and hard palate [6]. Osseous involvement indicates a T4a lesion. In patients with T4 oral cancers, the 5-year survival rate is close to 50%, whether treatment is surgical resection (47% survival) or chemotherapy (56% survival) [18]. Preoperative assessment of bone invasion in patients with oral cancer is essential for optimal surgical planning that is as noninvasive as possible. CT and MR imaging are complementary for assessing bone invasion, be it cortical and/or medullary, and may allow bone-sparing procedures in some cases [19]. The sensitivity of CT compared with histology is 40%–60%, with 89%–100% specificity, while MR imaging shows 56%–94% sensitivity and 73%–100% specificity [19]. The literature reported wide diagnostic variations for both.

CT is the standard imaging technique and most commonly used in determining bone involvement. Subtle cortical erosions are best detected using thin sectional CT (less than 3 mm slice thickness) images with bone algorithms using multiplanar reconstruction (Fig. 11). CT findings of osseous involvement include cortical erosion or interruption adjacent to the primary lesion, aggressive periosteal reaction, abnormal attenuation in bone marrow, and pathological fractures [2]. Although the positive predictive value of CT for mandibular invasion is considered satisfactory, its sensitivity remains inadequate for clinicians advocating intraoperative periosteal stripping when CT is negative [10].

Alternatively, the precise tumor extension with medullary bone involvement may be accurately assessed using MR imaging. MR findings suggestive of bone involvement include loss of hypointense cortex, replacement of hyperintense bone marrow by tumor on T1-weighted images, contrast enhancement within bones, and contrast enhancement of nerves traversing the mandible, especially the inferior alveolar nerve [2]. However, these imaging findings are unfortunately nonspecific, and conditions that may lead to false-positive findings of bone involvement include peritumoral edema, inflammation, coexisting periodontal disease, osteomyelitis, radiation-induced fibrosis, and osteoradionecrosis (Fig. 11) [2, 6]. Consequently, although MR imaging has good sensitivity and high negative predictive value for detecting bone involvement, its specificity is relatively low.

6.2 Perineural spread

Perineural spread of head and neck cancers is a characteristic form of metastatic disease and refers to macroscopic tumor spread beyond the main bulk of the primary tumor by traveling along the endoneurium or perineurium [10, 20]. It can be antegrade, or retrograde and skip lesions may occur; therefore,

72 Inflammation and oral cancer

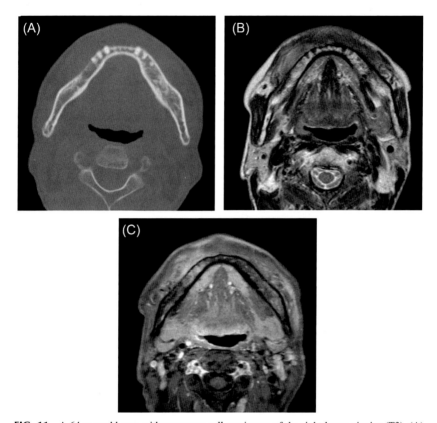

FIG. 11 A 64-year-old man with squamous cell carcinoma of the right lower gingiva (T3). (A) Unenhanced CT image with bone algorithm shows an erosion (*arrow*) of buccal cortex of the right mandible. (B) T2-weighted image shows a mild hyperintense lesion (*arrow*) of the right gingiva without bone marrow abnormality. (C) Fat-suppressed gadolinium-enhanced T1-weighted image shows mild enhancement (*arrow*) of the right mandible, which is a false-positive finding due to edema or inflammation.

meticulous evaluation of the skull base is important to detect perineural spread [6]. Although representative histologies are SCC and adenoid cystic carcinoma, other malignancies, including malignant lymphoma, and sarcoma can show perineural spread [21]. In oral cancers, the reported incidence of perineural spread varies from 5% to 10% [10]. Perineural spread is associated with poor prognosis and makes local and regional control particularly difficult by allowing tumor extension beyond the expected treatment margins. Because perineural invasion is often clinically silent and cannot be detected on physical examination, in particular, radiologists must be alert to this possibility [2].

Perineural spread of oral cancers is more common in subsites, including the floor of the mouth, where the neurovascular bundles are particularly accessible. In cancers of the floor of the mouth and oral tongue, perineural spread can course along the lingual nerve to the mandibular nerve and thence to the inferior alveolar nerve [10]. In hard palate cancers, perineural spread occurs through the greater and lesser palatine foramina and canal along the greater and lesser palatine nerves into the pterygopalatine fossa (Fig. 8) [10]. In retromolar trigone cancers, perineural spread can occur by the tumor extending posteriorly into the masticator space with further lateral spread to the mandibular foramina and thence along the inferior alveolar nerve [10]. In advanced gingival, buccal, and retromolar trigone cancers, perineural intracranial spread can occur along the mandibular nerve through the foramen ovale and contraindicates resection [10].

Although both CT and MR imaging can help detect perineural spread, MR imaging is the modality of choice because of its multiplanar capability, superior soft-tissue contrast, and the decreased number of metallic artifacts from dental fillings [20]. Disruption of the blood-nerve barrier leads to leakage of contrast agents, causing nerve enhancement before nerve enlargement [10]. CT features of perineural spread include replacement of normal fat, excessive enhancement, and abnormal enlargement of the nerve foramina. Foraminal widening or erosion is observed in more advanced cases (Fig. 8).

In addition, MR imaging can show perineural spread as the enlargement and enhancement of nerves and replacement of normal fat of the nerve foramina. Fat-suppressed contrast-enhanced T1-weighted images are particularly beneficial for detecting perineural spread. However, inflammatory or edematous enlargement of the nerves can simulate perineural spread; thus, these imaging findings are not specific for perineural spread [6]. Secondary signs of denervation, including atrophy of the muscles innervated by the involved nerve or abnormal enhancement of the denervated muscle, can help make the correct diagnosis [6]. Causes of denervation include direct involvement of the nerve by perineural spread and extrinsic compression or invasion by a neighboring tumor [22].

6.3 Lymph node metastasis

Because nodal involvement in patients with oral cancer is the single most important predictor, significantly affecting both treatment planning and patient survival, an accurate and simultaneous assessment of all nodal chains is required. A large percentage of oral cancers initially manifest with a neck mass representing the involvement of cervical nodes. However, the detection of nodal metastasis using imaging is more accurate than clinical examination; therefore, it has become routine to perform CT or MR imaging as workup [23]. Radiological identification of nodal disease is critical in surgical planning regarding neck dissection because imaging identifies metastasis in 7.5%–19% of clinically silent nodes [24].

The rates of regional metastasis from oral cancer vary according to the anatomical subsite and T category. Among oral cancers, those in the retromolar trigone and floor of the mouth are rich in lymphatics and show a strong

74 Inflammation and oral cancer

predilection for lymphatic involvement, with nearly 50% of patients presenting with regional disease [2]. In addition, SCCs in the oral tongue manifest with regional disease in 40% of patients, whereas SCCs in the lip, buccal mucosa, and hard palate are less likely to manifest with regional lymphadenopathy [2]. In patients with oral cancer, level I and II nodes are often the first to be involved. In previously untreated neck cancers, metastases to level IV and V nodes are rare in the absence of the involvement of level I–III nodes. Tongue cancers usually spread to ipsilateral level I and II nodes; however, skip metastases could occur to level III and IV nodes without the involvement of level I and II nodes, and contralateral metastases are known. Although extremely low rates (0.5%) of skip metastasis to level IV nodes have been reported in patients diagnosed with clinically N0 SCC of the oral cavity [25], a variable frequency of contralateral metastasis, ranging from 0.9% to 34.7%, has been reported in patients with SCC of the oral cavity [26]. Furthermore, cancers of the lip, floor of the mouth, gingiva, buccal mucosa, and retromolar trigone spread primarily to level IA, IB, and II nodes [10]. Radiologists should be familiar with the most probable sentinel nodes for a given malignancy [24].

Imaging assessment of cervical lymphadenopathy in patients with head and neck cancers includes the determination of size, morphological features, and margination. Nodal metastases are identified radiologically by larger nodal size, clustering, round shape, inhomogeneity, and extracapsular spread [24]. Among them, the cardinal imaging features of pathological lymphadenopathy are large nodal size and central necrosis within the node [2]. The usual size criterion is a maximal longitudinal diameter of more than 15 mm for jugulodigastric lymph nodes and more than 10 mm for other nodes (except for retropharyngeal lymph nodes, which are considered pathological at a maximal longitudinal diameter of more than 8 mm). However, if the minimal axial diameter is the parameter measured, a size of 11 mm for jugulodigastric (level II) nodes and 10 mm for other nodes is considered indicative of abnormality (except for retropharyngeal lymph nodes, which are considered pathological at a minimal axial diameter of more than 6 mm) [2, 27]. For a group of nodes in the drainage area of the primary tumor, the minimum axial diameter criterion decreases to 9 mm for level II nodes and 8 mm for other nodes [10]. Thus, size is the most frequently used criterion for diagnosis, although sensitivity and specificity vary widely, depending on the cutoff value used. However, the detection of nodal necrosis in patients with head and neck cancer is the most reliable sign of a metastatic node (Fig. 12) [28].

MR imaging is comparable to CT in terms of detecting necrosis [28]. Nodal necrosis may be identified as a central region of hypoattenuation with ring enhancement on CT images. If nodes appear with internal attenuation resembling that of water on CT images or with signal intensity of fluid on MR images, the term cystic adenopathy may be applied [2]. Necrosis should be differentiated from the fatty hilum of lymph nodes by measuring the CT attenuation and looking at the shape of the lymph node [10]. However, the detection of small nodes

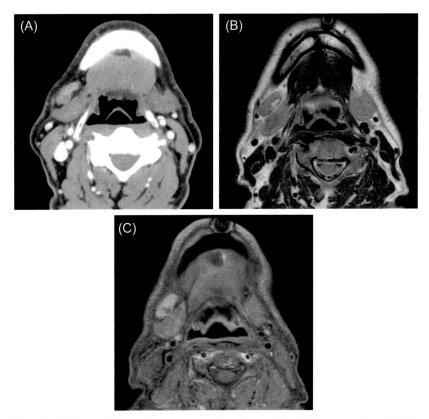

FIG. 12 A 76-year-old man with squamous cell carcinoma of the right lower gingiva with right submandibular lymph node metastasis. (A) Contrast-enhanced CT image shows an enlarged right submandibular lymph node (*arrow*) with focal necrosis (*arrowhead*). (B) T2-weighted image shows an enlarged right submandibular lymph node (*arrow*), but necrosis within the node cannot be detected. (C) Fat-suppressed gadolinium-enhanced T1-weighted image shows an enlarged right submandibular lymph node (*arrow*) with focal necrosis (*arrowhead*).

with micrometastases remains challenging for the currently available diagnostic imaging methods. Other morphological characteristics are equally important. Normal nodes tend to be reniform, whereas pathological nodes are more likely to be round [2]. CT and MR findings for pathological ENE include poorly defined margins, thickening or enhancement of the nodal rim, irregularity of the nodal capsule, and obliteration of the adjacent fat planes, whereas those for clinical ENE include skin thickening (skin invasion on physical examination), invasion to the musculature or deep layer in the deep cervical fascia (tethering to adjacent structures on physical examination), or nerve palsy as a secondary finding, such as vocal cord palsy, or muscle atrophy (invasion to the nerves with dysfunction on physical examination) [5].

6.4 Distant metastasis

Although distant metastases are rarely seen, if they occur, the lung is the most commonly affected site. ^{18}F-fluorodeoxyglucose PET/CT mainly plays a role in identifying regional lymphadenopathy and distant metastasis. However, because there is no additional value of PET/CT over CT or MR imaging in evaluating untreated clinically N0 SCCs of the oral cavity, the NCCN guidelines recommend the use of PET/CT only in stage III or IV cancers when management may be altered due to the detection of distant metastases [10]. If PET/CT is not employed, chest CT is used to rule out pulmonary metastases in higher T stage cancers and particularly when abnormal nodes are seen at level IV/supraclavicular regions [10]. Abdominal CT should be ordered only when the clinical index of hepatic metastasis is high [10].

7. Conclusion

CT and MR imaging play an important role in assessing tumor extension of primary tumors and regional nodal involvement. Sufficient knowledge of the characteristic anatomical subsites and potential routes of spread allows accurate staging in patients with oral cancer. These radiological imaging modalities help determine the appropriate staging, treatment strategy, and follow-up plan, and predict prognosis for patient morbidity and mortality.

References

[1] Palasz P, Adamski L, Gorska-Chrzastek M, Starzynska A, Studniarek M. Contemporary diagnostic imaging of oral squamous cell carcinoma—a review of literature. Pol J Radiol 2017;82:193–202.

[2] Trotta BM, Pease CS, Rasamny JJ, Raghavan P, Mukherjee S. Oral cavity and oropharyngeal squamous cell cancer: key imaging findings for staging and treatment planning. Radiographics 2011;31(2):339–54.

[3] Law CP, Chandra RV, Hoang JK, Phal PM. Imaging the oral cavity: key concepts for the radiologist. Br J Radiol 2011;84(1006):944–57.

[4] Kirsch C. Oral cavity cancer. Top Magn Reson Imaging 2007;18(4):269–80.

[5] Hiyama T, Kuno H, Nagaki T, et al. Extra-nodal extension in head and neck cancer: how radiologists can help staging and treatment planning. Jpn J Radiol 2020;38(6):489–506.

[6] Tshering Vogel DW, Zbaeren P, Thoeny HC. Cancer of the oral cavity and oropharynx. Cancer Imaging 2010;10:62–72.

[7] Chong V. Oral cavity cancer. Cancer Imaging 2005;5 Spec No A:S49–52.

[8] de Souza Figueiredo PT, Leite AF, Barra FR, et al. Contrast-enhanced CT and MRI for detecting neck metastasis of oral cancer: comparison between analyses performed by oral and medical radiologists. Dentomaxillofac Radiol 2012;41(5):396–404.

[9] Ong CK, Chong VF. Imaging of tongue carcinoma. Cancer Imaging 2006;6:186–93.

[10] Arya S, Rane P, Deshmukh A. Oral cavity squamous cell carcinoma: role of pretreatment imaging and its influence on management. Clin Radiol 2014;69(9):916–30.

Imaging findings of oral cancers **Chapter | 4** **77**

[11] Sarrion Perez MG, Bagan JV, Jimenez Y, Margaix M, Marzal C. Utility of imaging techniques in the diagnosis of oral cancer. J Craniomaxillofac Surg 2015;43(9):1880–94.

[12] Baba A, Hashimoto K, Kayama R, Yamauchi H, Ikeda K, Ojiri H. Radiological approach for the newly incorporated T staging factor, depth of invasion (DOI), of the oral tongue cancer in the 8th edition of American Joint Committee on Cancer (AJCC) staging manual: assessment of the necessity for elective neck dissection. Jpn J Radiol 2020;38(9):821–32.

[13] Sigal R, Zagdanski AM, Schwaab G, et al. CT and MR imaging of squamous cell carcinoma of the tongue and floor of the mouth. Radiographics 1996;16(4):787–810.

[14] Kimura Y, Sumi M, Sumi T, Ariji Y, Ariji E, Nakamura T. Deep extension from carcinoma arising from the gingiva: CT and MR imaging features. Am J Neuroradiol 2002;23(3):468–72.

[15] Imaizumi A, Yoshino N, Yamada I, et al. A potential pitfall of MR imaging for assessing mandibular invasion of squamous cell carcinoma in the oral cavity. Am J Neuroradiol 2006;27(1):114–22.

[16] Kato H, Kanematsu M, Makita H, et al. CT and MR imaging findings of palatal tumors. Eur J Radiol 2014;83(3):e137–46.

[17] Mukherji SK, Isaacs DL, Creager A, Shockley W, Weissler M, Armao D. CT detection of mandibular invasion by squamous cell carcinoma of the oral cavity. Am J Roentgenol 2001;177(1):237–43.

[18] Uribe S, Rojas LA, Rosas CF. Accuracy of imaging methods for detection of bone tissue invasion in patients with oral squamous cell carcinoma. Dentomaxillofac Radiol 2013;42(6), 20120346.

[19] Bouhir S, Mortuaire G, Dubrulle-Berthelot F, et al. Radiological assessment of mandibular invasion in squamous cell carcinoma of the oral cavity and oropharynx. Eur Ann Otorhinolaryngol Head Neck Dis 2019;136(5):361–6.

[20] Caldemeyer KS, Mathews VP, Righi PD, Smith RR. Imaging features and clinical significance of perineural spread or extension of head and neck tumors. Radiographics 1998;18(1):97–110 [quiz 147].

[21] Ojiri H. Perineural spread in head and neck malignancies. Radiat Med 2006;24(1):1–8.

[22] Borges A. Imaging of denervation in the head and neck. Eur J Radiol 2010;74(2):378–90.

[23] Hoang JK, Vanka J, Ludwig BJ, Glastonbury CM. Evaluation of cervical lymph nodes in head and neck cancer with CT and MRI: tips, traps, and a systematic approach. Am J Roentgenol 2013;200(1):W17–25.

[24] Dmytriw AA, El Beltagi A, Bartlett E, et al. CRISPS: a pictorial essay of an acronym to interpreting metastatic head and neck lymphadenopathy. Can Assoc Radiol J 2014;65(3):232–41.

[25] Warshavsky A, Rosen R, Nard-Carmel N, et al. Assessment of the rate of skip metastasis to neck level IV in patients with clinically node-negative neck oral cavity squamous cell carcinoma: a systematic review and meta-analysis. JAMA Otorhinolaryngol Head Neck Surg 2019;145(6):542–8.

[26] Gonzalez-Garcia R, Naval-Gias L, Sastre-Perez J, et al. Contralateral lymph neck node metastasis of primary squamous cell carcinoma of the tongue: a retrospective analytic study of 203 patients. Int J Oral Maxillofac Surg 2007;36(6):507–13.

[27] Zhang GY, Liu LZ, Wei WH, Deng YM, Li YZ, Liu XW. Radiologic criteria of retropharyngeal lymph node metastasis in nasopharyngeal carcinoma treated with radiation therapy. Radiology 2010;255(2):605–12.

[28] King AD, Tse GM, Ahuja AT, et al. Necrosis in metastatic neck nodes: diagnostic accuracy of CT, MR imaging, and US. Radiology 2004;230(3):720–6.

Chapter 5

What is epigenetics?

Keizo Kato
*Department of Oral and Maxillofacial Surgery, Gifu University Graduate School of Medicine,
Gifu, Japan*

1. Foreword

With the progress of molecular biology, life phenomena based on genetic interactions within living things, and various disease onset mechanisms triggered by regulation disorders thereof, have come to light. The transcription and translation of genes (genetics) is involved in the expression of proteins regulating the behavior of cells, and epigenetics is involved in the regulation of genetic transcription/translation. Furthermore, because epigenetics is controlled by proteins, the integrated control mechanisms of genetics-epigenetics-proteins involved in vital functions are believed to play a central role in life phenomena.

"Epigenetics" generally refers to changes of the inherited genetic expression or cell phenotype even after cellular division that are not accompanied by changes in the DNA nucleotide sequence [1]. It is related to cell generation, differentiation, and senescence as well as a variety of life phenomena, such as reprogramming. Epigenetics is involved in the ability of induced pluripotent stem cells and embryonic stem cells to become a variety of organs (differentiation), in factors influencing the success or failure and occurrence of abnormalities, etc., of mammalian clone creation (reprogramming), in the generation mechanisms for cancer and genetic diseases, and in brain function, among others.

Recent studies have revealed that, in addition to genetic abnormalities, cells accumulate various epigenetic abnormalities. Epigenetic abnormalities exert great influence on the characteristics of cancer, from the early stage of occurrence to its growth and progression. The effects of epigenetic abnormalities are thought to widely extend to genetic control abnormalities, and the elucidation of epigenetic abnormalities in cancer is a pressing issue in terms of cancer treatment. Epigenetic mechanisms regulate gene expression via the various forms of cross-talk of multiple epigenetic mechanisms, such as DNA methylation (a stabilized modification), histone modification (which maintains reversibility), chromatin structure change, and untranslated RNA. Recently, treatments targeting DNA methylation and histone modification abnormalities have been

Inflammation and Oral Cancer. https://doi.org/10.1016/B978-0-323-88526-3.00005-1
Copyright © 2022 Elsevier Inc. All rights reserved.

80 Inflammation and oral cancer

adopted in clinical settings, and it is expected that understanding the epigenetic control mechanisms in cancer cells will enable the elucidation of control mechanisms related to the onset and progression of cancer, and the prospect of an effective diagnosis and treatment targets.

The word "*epi*genetics" is created by fusing the Greek word "epi" (over/ beyond) with "genetics," and is used as a general term denoting the phenomena controlling genetic expression in the processes of ontogenesis and cell differentiation, without changing the genetic information written in the genome [1]. Subsequently, acquired changes of gene control not accompanied by changes of DNA nucleotide sequences have become a primary subject in the field of epigenetic research, and there have been a plethora of epigenetic studies since the 2000s, when the decoding of genomes had progressed for several animals.

The content of epigenetics as described above is not universally defined [2]. However, for an introductory explanation, it is often described as focusing on DNA methylation and the chemical modification of histones. The authors shall focus on DNA methylation in this chapter as, to date, the number of studies mainly centered around DNA methylation has been progressing.

For living things to develop and differentiate normally, it is necessary for the required genes to express at the proper timings, at a set amount specific to the tissues, and it is necessary to comprehend the expression control mechanisms for genetic information from the chromosomal genome. The methylation modification of DNA is the only method of directly modifying the mammalian genome, and performs control of gene expression by the attachment and removal of this methyl group. The methylation modification of DNA controls expression without changing the nucleotide sequence—that is, the gene itself—and without changing the amino acid sequence information coding the gene. In addition, once applied, the methylation pattern on the genome will be stably passed on to the next generation of cells. Furthermore, methylation modification also has as-needed removable reversibility, and how the methylation of genome DNA is regulated is important.

2. Methylation modification of DNA

On close observation, human genes have 500–2000 bp regions locally rich in CpG, and the cytosines of the CpG of these regions do not undergo methylation modification (CpG islands). Approximately 60% of the human housekeeping genes (the gene clusters coding the proteins necessary for cell survival) are thought to have CpG islands in the promoter regions. In the 1990s, the correlation of DNA methylation of the CpG islands of the Rb gene (a cancer controlling gene) with the suppression of gene expression was discovered. Thereafter, the correlation of DNA methylation and suppression of gene expression was reported at p16, a cancer-inhibiting gene, and the VHL gene and others were reported. At present, DNA methylation is recognized as a third mechanism, following deletion/mutation, among the gene deactivation mechanisms.

The connection between tumors and DNA methylation abnormalities was investigated for methylated cytosine content, and it was discovered that the DNA in tumors is generally in a hypomethylated state. The hypomethylation of tumors is thought to be triggered by the hypomethylation of repeated sequences in the genome. In other words, the DNA methylation in cancer cells is characterized by the hypomethylated state of the genome and hypermethylation in the CpG islands of specific genes, resulting in a state different to that of normal cells.

In cancer cells, multiple genes can be simultaneously suppressed by DNA methylation. The frequency of DNA methylation differs for each gene, and genes wherein the DNA methylation level increases with age and those that perform cancer-specific DNA methylation have been found. The targeted genes also differ according to the organ from which the cancer cells originate. Therefore, DNA methylation abnormalities are often thought of initially as a random phenomenon, but in actuality are selective rather than random. Since, according to a recent global analysis, high-frequency methylation of the CpG islands of genes not associated with the carcinogenesis process has not been discovered, genes beneficial to proliferation are not necessarily methylated.

When, among the DNA methylation target genes, genes inhibiting the infiltration and metastasis of cancer, such as E-cadherin (CDH1) and tissue inhibitor of metalloproteinase-3, are silenced, this results in cancers that have a high ability to metastasize and infiltrate. In addition, methylation of the O6-methylguanine DNA methyltransferase (DTMT; MGMT) gene triggers abnormalities in gene repair. Moreover, DNA methylation accumulates (CIMP: CpG island methylator phenotype) at a high frequency and exhibits a specific clinical profile in some cases of colon cancer, brain tumors (gliomas), and breast cancer, and the DNA methylation abnormalities are thought to affect the pathology of the cancer. Therefore, the detection of characteristics of DNA methylation abnormalities in cancer cells, and the further elucidation of the characteristic epigenetic profiles exerting an effect on the properties of cancer, are believed to be linked to the determination of individually adapted cancer treatment regimens.

In eukaryotes, most of the five cytosine bases (C) are subjected to methylation modification (Fig. 1). In animals, the number of genome bases increased and the cytosine methylation modification in the genome increased when vertebrates evolved. The genome DNA methylation of vertebrates continues with the cytosine/guanine bases and is overlaid to the cytosine bases in the CpG

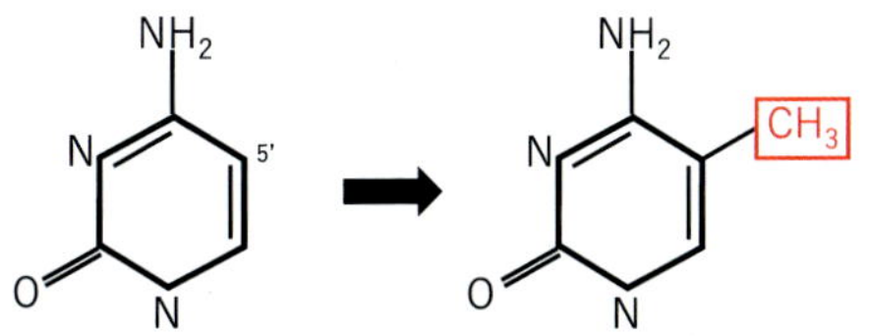

FIG. 1 Methylation modification of cytosine base.

82 Inflammation and oral cancer

sequence. This methyl group is transferred by the operation of DTMTs from S-adenosyl-L-methionine.

3. Writing with DNA methyltransferases

In vertebrates, the methylation states of gene regions are overwritten in the cell differentiation process during the reproductive cell maturation process and generation of an embryo following fertilization. In mammals, all methylation in reproductive cells is erased immediately and new methylation is written-in. Dnmt1, Dnmt3A, and Dnmt3B are DTMTs methylating DNA in mammals. The enzyme previously known as Dnmt2 is not a DTMT, and DNMT3L is believed to be involved in transcriptional repression without having methylation activity. Of these, Dnmt3A and Dnmt3B bring about new methylation in the genome. Once written, DNA methylation is transmitted to the genome of the daughter cell when DNA is replicated. This reaction is catalyzed by Dnmt1. Immediately after the methylated DNA has been synthesized, the methylation modifications remain on the parental side, but there are no methyl groups on the newly synthesized chains. Dnmt1 can methylate this state; thus, methylation is transmitted to the daughter cell (Fig. 2).

4. Reversibility of methylation

Although various possibilities have been suggested regarding the mechanism by which methylation is erased, the actual mechanism remains unknown. There are thought to be two mechanisms; one by which specific places are not methylated, and the other by erasing methyl groups bound to cytosine bases. If Dnmt1 is not functioning in the DNA replication process, half of somatic cells are demethylated by repeating division twice (Fig. 2). Other theories regarding demethylation include the excision repair of the methylated cytosine bases, and the methyl groups binding to the cytosine bases being directly excised by an enzyme. There are reports of both, but these are unclear. The demethylation reaction is an extremely important topic that requires elucidation.

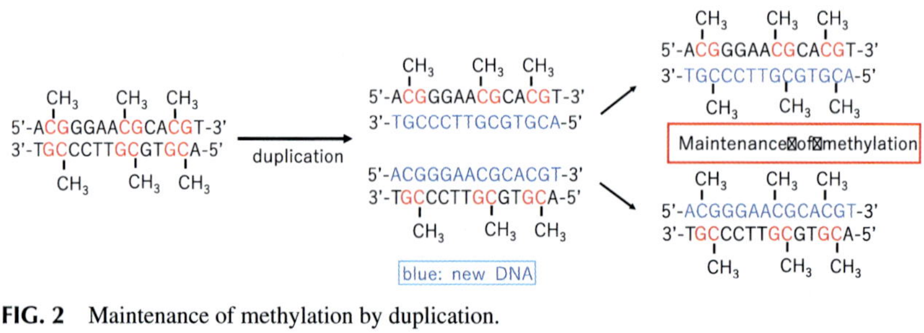

FIG. 2   Maintenance of methylation by duplication.

5. DNA methylation in cancer

Epigenetic abnormalities in cancer include abnormal DNA methylation and abnormal histone modification [3]. DNA hypermethylation and hypomethylation states have been proposed as DNA methylation abnormalities in cancer. DNA hypermethylation is involved in suppressing gene expression and hypomethylation in elevating gene expression. DNA methylation is an important regulating factor of gene transcription and hypermethylation or hypomethylation are found in many human malignancies, unlike in normal tissues. Hypomethylation is observed in a wide range of the genome and is thought to affect carcinogenesis through the destabilization of the genome and chromosomes. In most normal tissues, the quantity of methyl cytosine is 0.84% or higher, while in cancers this is decreased to below 0.84% [4]. Inside the genome, there is a wealth of repeating sequences in the CpG sequence and these are strongly methylated in normal tissues. In cancer tissues, the methylation of these repeating sequences frequently exhibits reduction.

However, the promoter regions of tumor suppressor genes that should be demethylated in normal cells often exhibit hypermethylation of the CpG islands of the gene promoters characteristic of abnormal DNA methylation in cancer cells, and gene deactivation is thought to be triggered by the repressed transcription of the hypermethylated genes to the mRNA. In addition, hypermethylation is frequently observed in CpG islands outside promoter regions and in regions outside the CpG islands, not just in the CpG islands of promoter regions.

6. CpG island standards

There is an often-used definition of "having a length of at least 200 base pairs, GC content of 50% or higher and the proportion of existing CpG is at least 60% of the amount expected from the GC content (CpG observed/expected > 0.6)" [5]. However, within the genome there exist Alu sequences, etc., of approximately 300 base pairs in length, in addition to sequences with high GC content. To exclude these, a new definition of "having a length of at least 500 base pairs, GC content of 55% or more, and CpG observed/expected > 0.65" has been advocated [6].

7. Activation of oncogenes

The relationship in humans between cancer and epigenetic effects on genes was first suggested in 1983 in a report of DNA hypomethylation occurring in colorectal cancer [7]. It was subsequently suggested that a cause/effect relationship exists between carcinogenesis and hypomethylation, as DNA hypomethylation is found in a variety of cancer tissues. Moreover, a hypothesis has been proposed that, because DNA hypomethylation activates gene expression, the expression of oncogenes is amplified by DNA

84 Inflammation and oral cancer

hypomethylation, leading to carcinogenesis. Although there have also been reports of hypomethylation of the H-RAS gene, a proto-oncogene, and others, the amplification of gene expression due to hypomethylation has not yet been proven.

8. Silencing of tumor suppressor genes

The deactivation of tumor suppressor genes is generated by the deactivation of both alleles through the combination of a mutation and chromosomal deficiency. Although the mechanism deactivating both alleles through mutation and chromosomal deficiency is well known, deactivation through methylation of the RB gene promoter region has been reported [8]. This proves that, having been hypermethylated, the gene expression of the RB genes, which are tumor suppressor genes, is suppressed and, as a result, the RB genes are quantitatively deactivated, generating the cancer. Various studies have subsequently been performed regarding methylation, and correlations between DNA methylation and the suppression of gene expression have been reported at the p16 and VHL genes, among others. At present, DNA methylation is recognized as the third mechanism among genetic deactivation mechanisms, after deficiency and mutation, and methylation is confirmed to be involved in the deactivation of many tumor suppressor genes.

On the other hand, the deactivation of tumor suppressor genes occurs with the deactivation of both alleles through the combination of a mutation and chromosomal deficiency. However, only alleles with methylated promoter regions have reduced expression due to methylation of the CpG islands of the promoter regions. The Knudson two-hit theory will be satisfied if gene polymorphisms, for example methylation and chromosomal deficiency, and methylation in both alleles, etc., are present in the promoter regions [9].

9. Causes of methylation abnormalities (Fig. 3)

9.1 Exogenous factors

Possible mechanisms for mutation induction are thought to include mutagenic substances in the environment, UV rays, tobacco, and alcohol. However, there are many unknowns regarding what mechanisms induce methylation.

9.2 Endogenous factors

Aging is an endogenous factor that induces methylation abnormalities. Cells with many methylated gene regions increase with age in ER genes. The most important sites of the CpG islands of promoter regions are protected by methylation, and are thought to have no effect on gene expression.

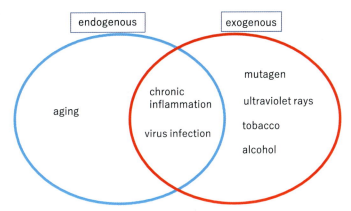

FIG. 3 Factors that cause abnormal DNA methylation.

10. Methylation detection methods (Fig. 4)

The accurate analysis of DNA methylation enables the understanding of the cancer and an accurate diagnosis. Methods for analyzing DNA methylation are roughly divided into two as follows: 1) methods analyzing individual gene regions; and 2) comprehensive genome analysis methods.

10.1 Analysis of individual gene regions

Methods analyzing individual gene regions include the bisulfite sequencing method, methylation-specific PCR (MSP), quantitative MSP, combined bisulfite restriction analysis (COBRA), and pyrosequencing, and their respective analysis ranges, sensitivities, accuracies, and necessary equipment differ considerably. All of these methods utilize the principle of converting the unmethylated tyrosine only to uracil through bisulfite processing.

The bisulfite sequencing method makes it possible to know the methylation state of each CpG in a specific region, but it is labor- and time-consuming. The MSP method is simple and the most widely used as it is sensitive enough to detect just 1 methylated molecule out of 1000, but it is not capable of quantitation. The quantitative MSP method uses a specific PCR primer on a methylated or unmethylated DNA template that has undergone bisulfite conversion processing, and measures the presence of DNA methylation modification according to the PCR amplification differential. It is capable of a wide range of quantitation, from less than 1% up to 100%, and the accuracy is high. The COBRA method is limited to restriction enzyme sequences, but is a simple quantitation method. The analysis equipment and reagents for pyrosequencing are expensive but it is capable of quantitating the methylation of individual CpGs with moderate and high sensitivity. It is also known as "sequencing by synthesis" and is capable of finding bisulfite-converted DNA in a specific range of concern. The 5mC level is determined by comparing the proportions of C and T bases at individual gene loci.

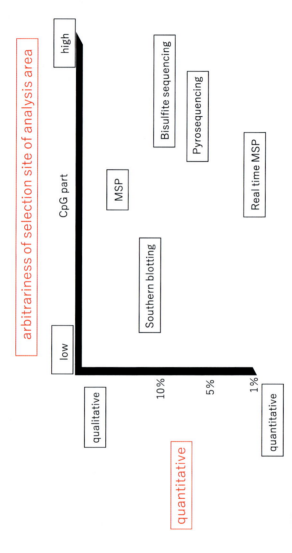

FIG. 4 Methylation detection methods.

10.2 Comprehensive genome analysis

The comprehensive genome DNA methylation analysis method is composed of two stages: the detection of DNA methylation and comprehensive screening of the genome. The detection of DNA methylation includes methods using affinity with antimethylated cytosine antibodies, etc., methods using methylation-sensitive restriction enzymes, methods with bisulfite processing, and expression-inducing methods using DNA demethylation agents. The current mainstream technique is to perform analysis of DNA regulated by any of the three methods described above through hybridization to a microarray or high-speed sequencing.

Comprehensive genome sequencing includes various microarray analyses, next-generation sequencing, two-dimensional electrophoresis, subtraction methods, etc. The resolution, quantitativity, cost-performance, and required equipment differ greatly depending on which combination is selected and must be used properly according to the respective research objective.

11. Methylation of various carcinomas

11.1 Oral cancer

Control factor genetics at cell cycle checkpoints or epigenetic changes occur in many tumors in the mouth. In particular, abnormalities of the spindle checkpoints acting on the mitotic stage (M stage) are involved in chromosomal instability and are thought to be important changes in canceration [10]. In squamous cell oral cancers, when this expression of the gene M stage checkpoint gene CHFR was searched for, expression is erased at a high rate through aberrant methylation, and in cell lines where CHFR expression is erased, G2 arrest due to docetaxel and paclitaxel administration does not occur and they exhibit high drug susceptibility. In cells where checkpoints due to CHFR do not work, this phenomenon is thought to have an action mechanism wherein the transit of cyclin-B1 to the nucleus cannot be blocked during mitotic cause stress. Therefore, it has been suggested that the presence of CHFR gene expression may be useful as a molecular marker of susceptibility to microtubule inhibitors, such as docetaxel and paclitaxel. In addition, those made to again express CHFR through treatment with AzaC, a methylation inhibitor, exhibit a markedly reduced mitotic index, suggesting the importance of checkpoints with CHFR in controlling entry into mitosis. In tumors where the M stage-regulating genes have been deactivated by methylation, abnormalities occur in the G2/M checkpoints, suggesting the high likelihood of susceptibility to anticancer drugs of types that impart DNA damage. Amplifying susceptibility to anticancer drugs of cancer cells in which the CHFR gene has not been methylated will become a future topic of interest [11].

CHFR abnormalities occur in a wide range of tumors, such as colorectal cancers, stomach cancers, leukemia, and oral cancers, and abnormalities occur

88 Inflammation and oral cancer

at the highest frequency in M stage checkpoint genes. Therefore, it is thought that the presence of CHFR methylation can be used as a marker for predicting susceptibility to microtubule inhibitors [11]. Regarding CHFR methylation in oral cancers, methylation has been observed at a frequency of 34.7% and, in addition to previous studies, this suggests that it is an important factor [12].

Kato et al. examined p16 and MGMT for the oral region as squamous cell cancers are common, just as in lung cancer. As a result, in oral cancer cases, p16 exhibited methylation at 50.9% and MGMT at 56.4% (Table 1). In addition, in normal tissues surrounding oral cancers, hypermethylation has been confirmed in p16 at 27.3% and MGMT at 40.9% (Table 2). These results suggest that hypermethylation in the oral region may occur as an early-stage event before canceration [13].

Epigenetic abnormalities typified in this manner by DNA methylation abnormalities have been reported and verified multiple times in various fields, suggesting involvement in carcinogenesis and cancer progression. In future, it will be important to analyze which genes are targets for the induction of DNA methylation abnormalities and their respective effects, and it is thought that combinations of gene methylations and gene mutations will provide clinically useful information not only for determining the carcinogenic mechanisms for various cancers but also for the individualization of treatment and prognosis prediction, etc.

TABLE 1 Promoter hypermethylation rate of p16 and MGMT genes in oral cancer and normal mucosa (healthy volunteer).

	p16			MGMT		
	Total	Methylation	Rate (%)	Total	Methylation	Rate
Malignant lesion						
Squamous cell carcinoma	51	25	49.02	51	27	52.94
Verrucous carcinoma	2	1	50	2	2	100
Carcinoma in situ	2	2	100	2	2	100
Total	55	28	50.9[a]	55	31	56.4[a]
Normal mucosa (healthy volunteer)	18	0	0	18	0	0

[a] *Significantly different from normal mucosa group by Fisher's exact test (P < 0.01).*

TABLE 2 Promoter hypermethylation rate of p16 and MGMT genes in OSCC and surrounding normal mucosa.

	p16			MGMT		
	Total	Methylation	Rale (%)	Total	Methylation	Rate (%)
Squamous cell carcinoma	22	16	72.73[a]	22	15	68.18 (NS)
Normal mucosa (surrounding)	22	6	27.27	22	9	40.91

[a] *Significantly different from normal mucosa (surrounding) by Fisher's exact test (P<0.05) NS not significantly different from normal mucosa (surrounding) by Fisher's exact test.*

11.2 Nonoral cancers

The involvement of the methylation of PML-RAR, the causative gene for promyelocytic leukemia, has been reported as a mechanism for suppressing the expression of the RARβ2 gene, thought to be one of the targets thereof [14]. PML-RAR is thought to combine with the promoter of the RARβ2 gene, inducing methylation by the DTMTs Dnmt1 and Dnmt3. In addition, 15 methylated genes have been newly discovered as examples of gene deactivation by the methylation of other gene promoter regions in human breast cancer by Miyamoto et al. [15]. One of the characteristics of breast cancer is the presence of many HER2 genes (HER2 gene amplification), and persons with such breast cancers have poorer prognoses than those with breast cancers with little HER2 gene amplification. When breast cancers are divided into three groups by quantity of DNA methylation abnormalities (cancer with many, few, and no abnormalities), HER2 gene amplification has been observed in cancers of the group with many DNA methylation abnormalities, and a link has been discovered between the quantity of DNA methylation abnormalities and HER2 gene amplification.

In addition, the correlation in ovarian cancer between the BRCA1 gene promoter methylation state of normal tissue with the risk of ovarian cancer is currently being investigated. As a result, in early-stage trials, the methylation of the BRCA1 gene was detected at a higher frequency in patients with high-malignancy serous ovarian cancer than in a control group (9.6% vs 4.2%, odds ratio 2.91). On the other hand, there is no difference between the control group and patients with low-malignancy serous ovarian cancer [16]. Furthermore, a lack of estrogen receptors in uterine cancer and BRCA1 deactivation in nonhereditary breast cancer can be demonstrated.

In addition, in dysplasia associated with ulcerative colitis, hypermethylation is observed in genes such as ER, Versican, and MyoD. The mean value of methylation of ulcerative colitis significantly increases in correlation with

90 Inflammation and oral cancer

the dysplasia-negative epithelium, low grade, high grade, and cellular variants, and suggests that part of the carcinogenic mechanism is borne by chronic inflammation.

A cohort study has been reported wherein accumulation of tumor-specific promoter CGI hypermethylation was observed upon analysis of gene methylation in colorectal cancer, and the concept of the CIMP has been put forward [17, 18]. Although the details of the molecular mechanisms of CIMP-positive colorectal cancer have not been elucidated, these tumors have a high rate of BRAF and KRAS gene mutations and are characterized by a pathology exhibiting geriatric onset, high incidence in the deep large bowel, mucus production, and a low differentiation tendency. CIMP-positive colorectal cancer is further classified into two molecular sub-types, CIMP1 and CIMP2, suggesting occurrence by different carcinogenic routes from their respective specific genetic abnormalities. Even among CIMP-positive cases, CIPMP1, which has a higher degree of gene methylation, exhibits epigenetic silencing of the mismatch repair gene MLH1, microsatellite instability (MSI), and BRAF gene mutations at high frequency [18]. In CIMP1 colorectal cancer, high frequency mutations of chromatin remodeling genes, starting with CHD7 and CHD8, are seen upon analysis by exome sequencing [19]. CIMP2 colon cancer exhibits a lower degree of gene methylation compared to CIMP1 colorectal cancer, and a high frequency of KRAS gene mutations. On the other hand, the frequency of methylated genes is low in CIMP-negative colorectal cancer, and TP53 gene mutation and chromosome instability are seen at a high frequency [17, 18]. Even among adenomas, which are colorectal cancer lesions, large bowel serrated adenomas tend to occur in the deep large bowel and exhibit BRAF gene mutations and CIMP traits thus, these may be precancerous lesions for CIMP1 colorectal cancer, and CIMP-positive colorectal cancer may occur from precancerous lesions that differ to normal. Furthermore, the methylation of the MLH1 gene that codes the mismatch repair protein exhibits a strong correlation with MSI in CIMP-positive colorectal cancer [18].

12. Possibility of application in cancer treatment

Epigenetic abnormalities are reversible reactions, and the expression of genes deactivated by the hypermethylation of promoters can be recovered with DNA methylation inhibitors. This is a great difference from genetic abnormalities.

The application of methylation inhibitors in cancer treatment has a long history, and therapeutic studies involving antileukemia drugs began in the 1960s. Their antitumor effects at low concentrations with few side effects have recently been reported, and a future therapeutic effect in combination with other chemotherapy agents is anticipated. Lubbert et al. reported a response rate of 60% with the administration of the methylation inhibitor 5-aza-deoxycytidine (DAC) to MDS cases, and in the DAC-responsive cases, expression induction through p15INK4a demethylation was observed [20]. Moreover, Issa et al. reported a response rate of 60% with the administration of DAC in acute myelocytic leukemia [21].

There have also been attempts to utilize epigenetic abnormalities as markers of drug susceptibility. Esteller et al. found that genes are deactivated by the aberrant methylation of DNA by the DNA recovery enzyme MGMT. Tumor cell lines methylated by MGMT are notable for their high susceptibility to alkylating drugs, and when the efficacies of BCNU (Gliadel wafer) in a MGMT methylated and a nonmethylated group were compared, there was a significant effect in the methylated group [22].These results suggest that treatment methods can be selected using the biological characteristics of the tumor cells specified by methylation, which is thought to be promising with regard to future treatment applications.

Although it had been previously indicated that the overall methylation of cancer cells is lower compared to that of normal cells, it was discovered that in renal cancer, the methylation of the cancer suppressor gene VHL and the CpG islands of the p16 promoter regions is hyperactive compared to that of normal cells, while simultaneously inhibiting the expression of these genes. The genes that are subject to cancer-specific deactivation through hypermethylation include many genes for the cell cycle, apoptosis, signal transmission, DNA repair, etc., and not just the previously known cancer suppressor genes. Until now, it was thought that deactivation of cancer suppressor genes and oncogenes occurs in cancer cells that exhibit hypomethylation of the genome as a whole but that exhibit partial hypermethylation. Although the reasons why such aberrant methylation occurs are largely unknown, hypermethylation of cancer suppressor genes and oncogenes occurs in various cancers with the hypermethylation of some genes exhibiting organ specificity. Furthermore, multiple genes exhibit hypermethylation, thus underlining the reality of their usefulness in the early discovery and treatment of cancer.

13. Hypermethylation detection in blood and body fluids and assay methods for clinical application

Various studies have been performed using sputum, blood, bronchial lavage fluid, etc., and Belinsky et al. have reported on the methylation of p16 with regard to lung cancer and the sputum of patients with lung cancer [23]. The frequency of methylation increases not just in the squamous cell carcinoma of the lungs, but also as the condition progresses from hyperplasia to metaplasia, and to carcinoma in situ [23]. Furthermore, p16 methylation is also detected from not only the sputum of patients with lung cancer, but also the sputum of smokers, who are a high-risk group for the onset of lung cancer.

Palmissano et al. [24] conducted an investigation after performing sputum collection prior to definitive diagnosis on lung cancer high-risk group subjects, such as smokers and persons with experience of working with radon ores, etc. As a result, by targeting p16 and MGMT hypermethylation, gene

92 Inflammation and oral cancer

hypermethylation was detected from the sputum of almost 100% of lung cancer patients. Furthermore, the sputum prior to definitive diagnosis exhibited hypermethylation of either p16 or MGMT. The hypermethylation of cancer suppressor genes displays effectiveness not only as a genetic marker for the early discovery of cancer, but also for the screening of high-risk groups for cancer onset.

In breast cancer, upon investigating the hypermethylation of CycD2, RARβ, and Twist from the breast aspirate of 56 high-risk individuals, hypermethylation was observed in 37% of cases exhibiting a high degree of atypicality in cytodiagnosis [25].

In addition, Nagata et al. [26] searched for methylation using mouthwash as a specimen. As a result, methylation was observed in E-CDH, TMEFF2, RARβ, and MGMT. By combining these four genes, they discovered a new and noninvasive method of detecting oral cancers with a sensitivity and specificity of over 90%, and suggested the possibility of a future diagnostic tool for oral squamous cell carcinomas.

These studies involved the collection of blood, plasma, body fluids, etc., by the collection of live samples through noninvasive methods, also called "liquid biopsies." By improving the research tools and analysis methods, it has become possible to implement genetic tests and epigenetic tests on trace amounts of substances present in general liquid biopsy samples. These new tests may provide a lot of information more quickly than previous approaches, enabling faster diagnosis and better treatment strategies. Furthermore, because liquid biopsy is less costly and invasive than tissue biopsy, it is thought to be superior in placing fewer burdens on patients.

In addition, methods are also being developed that are capable of methylation assays that are simpler than conventional methods. Normally, DNA is formed in a double-helix structure, but DNA with specific sequences has a quadruplex structure. When quadruplex DNA is methylated, the PCR amplification efficiency is markedly reduced when amplified by PCR. As a result of having measured the amplification efficiency of quadruplex regions in the oncogene VEGF-A by real-time PCR with the human genome as the subject, a clear methylation frequency-dependent decrease was found, suggesting the possibility of easy measurements of methylation frequency of target oncogenes [27].

These noninvasive test and assay methods are expected to bring about major future developments in the field of epigenetics centered on methylation and to contribute to the identification of factors related to cancer patient prognosis, drug reactivity, and carcinogenesis risk that could not be revealed by conventional clinical factors alone.

14. Toward the realization of methylation therapies

Epigenetic pharmaceuticals are possible alternatives or adjuvant therapies for currently received treatment methods, such as radiotherapy and chemotherapy, and recent studies have shown that they are likely able to enhance the effect of current therapies [28]. In addition, it has been suggested that methylation is connected to treatment and prognosis, not just early diagnosis. Esteller et al. reported on MGMT hypermethylation and the prognosis of treatment with alkylating agents [22]. In gliomas, MGMT hypermethylation indicates a good prognosis, indicating that methylation is important as a prognostic factor.

Myelodysplastic syndrome (MDS) is one of the major hematopoietic tumors together with leukemia, malignant lymphoma, and multiple myeloma. It is characterized by approximately 30% transitioning to acute myeloid leukemia (AML), the prognosis of which is poor as the morbidity period is longer. IPSS classification is performed based on chromosomal testing and placed in four risk categories as follows: high-risk and low-risk, with intermediate 1 and intermediate 2 in-between these. Although the median survival time for low risk is 5.7 years, for high risk it is extremely short at 0.4 years. Genetic mutations are multifarious, and although genetic point mutations or deletions are also conceivable, it is thought almost certain that translocations occur because of the increase in chronic myeloid leukemia. Furthermore, it is possible that mutations of the genes pertaining to DNA methylation control that have recently attracted attention are included. The reason that a relationship with DNA methylation is hypothesized is because it has become apparent that azacytidine, which is believed to be the main mechanism of the demethylation, controls the transition from MDS to AML, improving patient prognosis. Azacytidine is integrated to the DNA synthesized by cells during proliferation because its structure is similar to that of nucleic acids. It is thought that upon integration into DNA, it reversibly binds with DTMT, the enzyme catalyzing the DNA methylation reaction, exhausting the free DTMT, resulting in the DNA being in a hypomethylated state and, as a result, releasing the methylation of the cancer suppressor genes, resulting in detumorization.

15. Oral cancer treatment

As per the above, hypermethylation is a major target of cancer treatment. The demethylation agent DAC has a strong DNA methylation inhibition action and has a growth inhibition effect on tumor cells. Treatments on hematologic malignancies, mainly leukemia, are currently being investigated [29].

The cytotoxic mechanisms with DAC are thought to be of the two following types: (1) a direct inhibitory action on DTMT. In sputum, DTMT 1 (Dnmt1) binds with Rb proteins and histone deacetylase (HDAC), etc., without overlying the methyl groups in the DNA, is involved in the cell cycle and DNA replication phase, and is thought to be strongly correlated to the cell growth inhibitory effect. (2) Expression recovery by the demethylation of genes with a growth inhibition action, such as hypermethylated cancer suppressor genes, and a correlation is exhibited between the expression recovery of p16 with DAC and the growth inhibition. However, DAC requires high concentrations for the demethylation action and there are concerns regarding side effects, such as myelosuppression. Therefore, we are currently focusing on epigallocatechin gallate (EGCG) as a substance transformed to DAC. EGCG is a natural substance with a demethylation action and is contained in large amounts mainly in green tea (Japanese tea). At normal intake amounts, it is a safe substance without adverse effects on the body.

In a study that used an oral cancer cell line with methylation in the RECK gene, EGCG was proven to have a demethylation action, cancer infiltration suppression ability, and cell migration inhibition on par with DAC [30]. The RECK gene is involved in the infiltration/metastasis of tumors by inhibiting the activity of three types of matrix metalloproteinases (MMPs): MMP-2, MMP-9, and MT1-MMP (Fig. 5). The hyperactive expression of any of these three MMPs accompanying cancer progress is clear, and these MMPs are involved in cancer infiltration/metastasis and tumor angiogenesis as collagenolytic enzymes. RECK is located on the surfaces of normal cells and simultaneously controls the activity of these MMPs. Therefore, when oncogenes reduce RECK expression, these MMPs are activated at once, accumulate reduced expression of the extracellular matrix proteins and the receptors (integrins) thereof, weakening cell-matrix adhesion, and are surmised to play a part in malignant trait expression. Our studies applying EGCG found a high degree of hypermethylation of RECK promoters in oral squamous cell carcinoma cell lines and were the first to show a correlation with reduced mRNA expression. Furthermore, the green tea polyphenol EGCG reversed hypermethylation and could amplify RECK mRNA expression. ECGC significantly inhibited cancer cell infiltration capacity in a three-dimensional collagen infiltration model through the demethylation effect on MMP inhibitor genes or the direct inhibition of MMP activity. EGCG is significant as a powerful natural food for chemotherapy and chemotherapeutic approaches in oral cancer and is thought to have benefits in opening up future prospects outside of medications (Figs. 6–9, Table 3).

In addition, although they are not DNA methylation inhibitors, clinical applications, and studies regarding histones are underway. The clinically applied medications among HDAC inhibitors include valproic acid and suberoylanilide hydroxamic acid, and treatments using these medications are currently being used, centered on hematologic malignancies. HDAC inhibitors stop the cell cycle by inducing p21 expression and activating p53 and are thought to show antitumor action through the induction of apoptosis. In addition, they are connected to the reactivation of multiple genes through the acetylation of deactivated genes.

The concomitant use of HDAC inhibitors with other molecularly targeted therapeutic agents is also being investigated. Although differentiation induction therapies with all-trans retinoic acid (ATRA), a vitamin A preparation, are often effective for acute promyelocytic leukemia (APL), ATRA therapy-resistant APL also occurs sometimes. A previous study induced differentiation in such APL by the combined use of HDAC inhibitors and ATRA [31]. Nonsmall cell lung cancer, in which there is an epithelial growth factor receptor (EGFR) mutation,

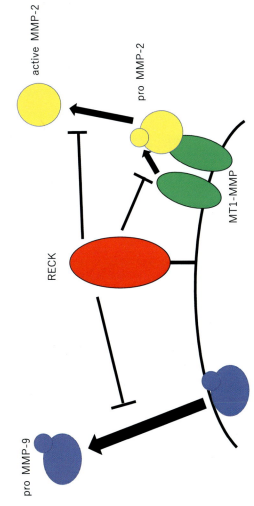

FIG. 5 RECK controls three MMP family molecules.

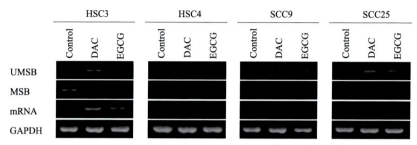

FIG. 6 Alterations of methylation status and mRNA expression levels of RECK gene in OSCC cell lines after treatment with 50 μM EGCG or 8.7 μM 5-aza-dC(DAC) for 6 days. GAPDH was used as internal control.

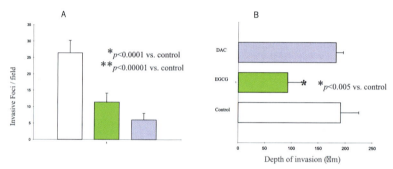

FIG. 7 Invasive foci formation and invasion depth of cancer cells were inhibited by EGCG. (A) EGCG or 5-aza-dC significantly blocked cancer invasion in HSC3 cells by decreasing the mean number of invasive foci/field by t-test. Results are shown as the mean number of invasive foci ± 1SD in five randomly selected fields (6 days). (B) 6 days after addition of HSC3 cells to gels, the depths of cell invasion of cells that had invaded the gel were significant blocked in cells treated with EGCG.

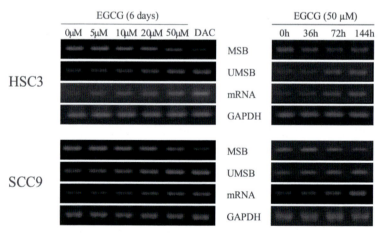

FIG. 8 Dose-dependent and time course study. SCC9 and HSC3 cells were treated with 5, 10, 20, or 50 μM EGCG, or 8.7 μM 5-aza-dC for 6 days. Cell lines were treated for 32, 72, and 144 h with 50 μM of EGCG.

What is epigenetics? Chapter | 5 97

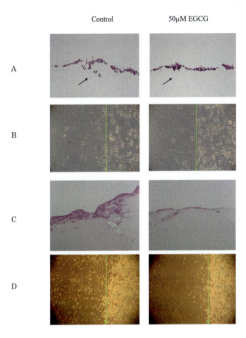

FIG. 9 Collagen gel invasion model. Treatment of EGCG for 6 days inhibits cancer invasion in collagen model in HSC3 cells. (A) The collagen gel was collected and fixed in 10% formalin, embedded in paraffin, stained with hematoxylin and eosin, and examined for cancer cell invasion. Treatment with EGCG significantly inhibits invasion in collagen model for 6 days (*arrow* is showing an invasive foci). (B) Cell migration ability was suppressed in presence of EGCG for 6 days. Cells atop collagen gels with surfaces were partially covered by glass cover slips (1×2 cm). When the cells reach confluence, the glass fragments were removed, leaving a cell-free area on the gel. The distances migrated across the gels were observed 24 h later using inverted microscope. (C) Inhibition of cell invasion by EGCG in HSC3 cells at 14 days. (D) The ability of cancer cells to migrate from the edge toward center of the uncovered gel in HSC3 cells cultured with EGCG were decreased as compared to control on day 14 (after 24 h of removing the cover slips) (×100 magnification). The arrows show the margin of the uncovered area gel.

is generally highly susceptible to tyrosine kinase inhibitors (TKIs). As a result of having performed combined administration of HDAC inhibitors and TKIs on EGFR-mutant lung cancer cell lines that exhibit resistance to TKIs, recovery of TKI susceptibility and growth inhibition effects have become visible [32].

Clinical applications of histone methylase inhibitors are also progressing. Although 3-deazaneplanocin A (DZNep) was originally developed as an antiviral, there are reports that it can inhibit H3K27 trimethylation by inhibiting PRC2, which includes EZH2, leading to the apoptosis of cancer cells [33]. Suva et al. reported that DZNep causes morphological changes in cancer stem cells of brain tumors and brings about growth inhibition [34]. EZH2 inhibitors have been identified in the USA, and deployment for clinical applications is progressing rapidly.

98 Inflammation and oral cancer

TABLE 3 Mean number of invasive foci per field and depth of invasion in HSC3 and HSC4 cancer cell lines after treatment with 50 µM EGCG or 8.7 µM 5-aza-dC(DAC) for 6 days.

Cell lines	Number of invasive foci per field Mean ± SD			Depth of invasion Mean ± SD (µm)		
	Control	50 µM EGCG	8.7 µM DAC	Control	50 µM EGCG	8.7 µM DAC
HSC3	26.4±3.8	11.4±2.7[a]	6.0±2.0[b]	191.0±33.7	93.0±32.3[c]	182.4±13.2
HSC4	17.6±2.9	10.2±0.8[d]	10.6±3.6[c]	234.6±27.9	180.4±16.1[a]	206.4±25.9

[a] Significantly different from control by the Student's t-test (P<0.0001).
[b] Significantly different from control by the Student's t-test (P<0.00001).
[c] Significantly different from control by the Student's t-test (P<0.005).
[d] Significantly different from control by the Student's t-test (P<0.001).

16. Future perspectives

In recent years, molecular drugs targeted to specific molecules related to carcinogenesis have been developed in cancer therapies, and these have exhibited superb effects. On the other hand, therapeutic agents targeted to epigenetics are also being clinically applied. Effects have been observed on some tumors, and these are becoming a new option for cancer treatment strategies. However, several issues remain in epigenetic treatments for cancer. For example, epigenetic therapeutic agents called "demethylation agents" and "histone deacetylase inhibitors" have a direct action inhibiting modifying enzymes, and their action mechanisms are clear, similar to other molecularly-targeted drugs; however, there are many unknowns regarding their downstream targets. In addition, the types of therapeutic agents are limited for a variety of epigenetic mechanisms, and the development of new therapeutic agents is desired. With the development of epigenetic therapies as true molecularly targeted drugs, how to increase specificity and on which diseases to perform epigenetic therapies are important future topics. The deployment of studies on the epigenetic abnormalities of individual cancers is anticipated in terms of the selection of more effective therapies.

References

[1] Egger G, Liang G, Aparicio A, Jones PA. Epigenetics in human disease and prospects for epigenetic therapy. Nature 2004;429(6990):457–63.

[2] Ledford H. Disputed definitions. Nature 2008;455(7216):1023–8.

[3] Jones PA, Baylin SB. The fundamental role of epigenetic events in cancer. Nat Rev Genet 2002;3(6):415–28.

What is epigenetics? Chapter | 5 **99**

[4] Gama-Sosa MA, Slagel VA, Trewyn RW, Oxenhandler R, Kuo KC, Gehrke CW, Ehrlich M. The 5-methylcytosine content of DNA from human tumors. Nucleic Acids Res 1983;11(19):6883–94.

[5] Gardiner-Garden M, Frommer M. CpG islands in vertebrate genomes. J Mol Biol 1987;196(2):261–82.

[6] Takai D, Jones PA. Comprehensive analysis of CpG islands in human chromosomes 21 and 22. Proc Natl Acad Sci U S A 2002;99(6):3740–5.

[7] Feinberg AP, Vogelstein B. Hypomethylation distinguishes genes of some human cancers from their normal counterparts. Nature 1983;301(5895):89–92.

[8] Ohtani-Fujita N, Fujita T, Aoike A, Osifchin NE, Robbins PD, Sakai T. CpG methylation inactivates the promoter activity of the human retinoblastoma tumor-suppressor gene. Oncogene 1993;8(4):1063–7.

[9] Jones PA, Laird PW. Cancer epigenetics comes of age. Nat Genet 1999;21(2):163–7.

[10] Toyota M, Sasaki Y, Satoh A, Ogi K, Kikuchi T, Suzuki H, Mita H, Tanaka N, Itoh F, Issa J-PJ, Jair K-W, Schuebel KE, Imai K, Tokino T. Epigenetic inactivation of CHFR in human tumors. Proc Natl Acad Sci U S A 2003;100(13):7818–23.

[11] Ogi K, Toyota M, Mita H, Satoh A, Kashima L, Sasaki Y, Suzuki H, Akino K, Nishikawa N, Noguchi M, Shinomura Y, Imai K, Hiratsuka H, Tokino T. Small interfering RNA-induced CHFR silencing sensitizes oral squamous cell cancer cells to microtubule inhibitors. Cancer Biol Ther 2005;4(7):773–80.

[12] Baba S, Hara A, Kato K, Long NK, Hatano Y, Kimura M, Okano Y, Yamada Y, Shibata T. Aberrant promoter hypermethylation of the CHFR gene in oral squamous cell carcinomas. Oncol Rep 2009;22(5):1173–9.

[13] Kato K, Hara A, Kuno T, Mori H, Yamashita T, Toida M, Shibata T. Aberrant promoter hypermethylation of p16 and MGMT genes in oral squamous cell carcinomas and the surrounding normal mucosa. J Cancer Res Clin Oncol 2006;132(11):735–43.

[14] Di Croce L, Raker VA, Corsaro M, Fazi F, Fanelli M, Faretta M, Fuks F, Coco FL, Kouzarides T, Nervi C, Minucci S, Pelicci PG. Methyltransferase recruitment and DNA hypermethylation of target promoters by an oncogenic transcription factor. Science 2002;295(5557):1079–82.

[15] Miyamoto K, Fukutomi T, Akashi-Tanaka S, Hasegawa T, Asahara T, Sugimura T, Ushijima T. Identification of 20 genes aberrantly methylated in human breast cancers. Int J Cancer 2005;116(3):407–14.

[16] Lønning PE, Berge EO, Bjørnslett M, Minsaas L, Chrisanthar R, Høberg-Vetti H, Dulary C, Busato F, Bjørneklett S, Eriksen C, Kopperud R, Axcrona U, Davidson B, Bjørge L, Evans G, Howell A, Salvesen HB, Janszky I, Hveem K, Romundstad PR, Vatten LJ, Tost J, Dørum A, Knappskog S. White blood cell BRCA1 promoter methylation status and ovarian cancer risk. Ann Intern Med 2018;168(5):326–34.

[17] Toyota M, Ahuja N, Ohe-Toyota M, Herman JG, Baylin SB, Issa JP. CpG island methylator phenotype in colorectal cancer. Proc Natl Acad Sci U S A 1999;96(15):8681–6.

[18] Shen L, Toyota M, Kondo Y, Lin E, Zhang L, Guo Y, Hernandez NS, Chen X, Ahmed S, Konishi K, Hamilton SR, Issa JP. Integrated genetic and epigenetic analysis identifies three different subclasses of colon cancer. Proc Natl Acad Sci U S A 2007;104(47):18654–9.

[19] Tahara T, Yamamoto E, Madireddi P, Suzuki H, Maruyama R, Chung W, Garriga J, Jelinek J, Ya-mano HO, Sugai T, Kondo Y, Toyota M, Issa JP, Estecio MR. Colorectal carcinomas with CpG island methylator phenotype 1 frequently contain mutations in chromatin regulators. Gastroenterology 2014;146(2):530–538.e5.

[20] Daskalakis M, Nguyen TT, Nguyen C, Guldberg P, Köhler G, Wijermans P, Jones PA, Lübbert M. Demethylation of a hypermethylated P15/INK4B gene in patients with myelodysplastic syndrome by 5-aza-2'-deoxycytidine (decitabine) treatment. Blood 2002;100(8):2957–64.

100 Inflammation and oral cancer

[21] Issa JP, Garcia-Manero G, Giles FJ, Mannari R, Thomas D, Faderl S, Bayar E, Lyons J, Rosenfeld CS, Cortes J, Kantarjian HM. Phase 1 study of low-dose prolonged exposure schedules of the hypomethylating agent 5-aza-2′-deoxycytidine (decitabine) in hematopoietic malignancies. Blood 2004;103(5):1635–40.

[22] Esteller M, Garcia-Foncillas J, Andion E, Goodman SN, Hidalgo OF, Vanaclocha V, Baylin SB, Herman JG. Inactivation of the DNA-repair gene MGMT and the clinical response of gliomas to alkylating agents. N Engl J Med 2000;343(19):1350–4.

[23] Belinsky SA, Nikula KJ, Palmisano WA, Michels R, Saccomanno G, Gabrielson E, Baylin SB, Herman JG. Aberrant methylation of p16(INK4a) is an early event in lung cancer and a potential biomarker for early diagnosis. Proc Natl Acad Sci U S A 1998;95(20):11891–6.

[24] Palmisano WA, Divine KK, Saccomanno G, Gilliland FD, Baylin SB, Herman JG, Belinsky SA. Predicting lung cancer by detecting aberrant promoter methylation in sputum. Cancer Res 2000;60(21):5954–8.

[25] Evron E, Dooley WC, Umbricht CB, Rosenthal D, Sacchi N, Gabrielson E, Soito AB, Hung DT, Ljung B, Davidson NE, Sukumar S. Detection of breast cancer cells in ductal lavage fluid by methylation-specific PCR. Lancet 2001;357(9265):1335–6.

[26] Nagata S, Hamada T, Yamada N, Yokoyama S, Kitamoto S, Kanmura Y, Nomura M, Kamikawa Y, Yonezawa S, Sugihara K. Aberrant DNA methylation of tumor-related genes in oral rinse: a noninvasive method for detection of oral squamous cell carcinoma. Cancer 2012;118(17):4298–308.

[27] Yoshida W, Yoshioka H, Bay DH, Iida K, Ikebukuro K, Nagasawa K, Karube I. Detection of DNA methylation of G-quadruplex and i-motif-forming sequences by measuring the initial elongation efficiency of polymerase chain reaction. Anal Chem 2016;88(14):7101–7.

[28] Wang LG, Chiao JW. Prostate cancer chemopreventive activity of phenethyl isothiocyanate through epigenetic regulation (review). Int J Oncol 2010;37(3):533–9.

[29] Bohl SR, Bullinger L, Rücker FG. Epigenetic therapy: azacytidine and decitabine in acute myeloid leukemia. Expert Rev Hematol 2018;11(5):361–71.

[30] Kato K, Long NK, Makita H, Toida M, Yamashita T, Hatakeyama D, Hara A, Mori H, Shibata T. Effects of green tea polyphenol on methylation status of RECK gene and cancer cell invasion in oral squamous cell carcinoma cells. Br J Cancer 2008;99(4):647–54.

[31] Kitamura K, Hoshi S, Koike M, Kiyoi H, Saito H, Naoe T. Histone deacetylase inhibitor but not arsenic trioxide differentiates acute promyelocytic leukaemia cells with t(11;17) in combination with all-trans retinoic acid. Br J Haematol 2000;108(4):696–702.

[32] Sharma SV, Lee DY, Li B, Quinlan MP, Takahashi F, Maheswaran S, McDermott U, Azizian N, Zou L, Fischbach MA, Wong KK, Brandstetter K, Wittner B, Ramaswamy S, Classon M, Settleman J. A chromatin-mediated reversible drug-tolerant state in cancer cell subpopulations. Cell 2010;141(1):69–80.

[33] Tan J, Yang X, Li Z, Jiang X, Chen W, Lee PL, Karuturi RK, Tan PB, Liu ET, Yu Q. Pharmacologic disruption of Polycomb-repressive complex 2-mediated gene repression selectively induces apoptosis in cancer cells. Genes Dev 2007;21(9):1050–63.

[34] Suvà ML, Riggi N, Janiszewska M, Radovanovic I, Provero P, Stehle JC, Baumer K, Le Bitoux MA, Marino D, Cironi L, Marquez VE, Clément V, Stamenkovic I. EZH2 is essential for glioblastoma cancer stem cell maintenance. Cancer Res 2009;69(24):9211–8.

Chapter 6

Role of autophagy in dysregulation of oral mucosal homeostasis

Madoka Yasunaga[a], Masahiro Yamaguchi[b], Kei Seno[c], Mizuki Yoshida[b], and Jun Ohno[d]

[a]Section of Orthodontics, Department of Oral Growth and Development, Fukuoka Dental College, Fukuoka, Japan, [b]Section of Gerodontology, Department of General Dentistry, Fukuoka Dental College, Fukuoka, Japan, [c]Section of General Dentistry, Department of General Dentistry, Fukuoka Dental College, Fukuoka, Japan, [d]Oral Medicine Research Center, Fukuoka Dental College, Fukuoka, Japan

1. Introduction

Establishing and maintaining oral mucosal homeostasis is essential for human health [1]. The oral epithelium helps maintain oral mucosal homeostasis as a physical barrier, protecting the underlying tissues from microbial invasion and mechanical forces. Cellular homeostasis depends on a balance between the production and destruction of macromolecules and organelles. In circumstances when cells cannot turnover damaged organelles or remove misfiled protein aggregates and foreign organisms, dysregulation of homeostasis occurs, resulting in a wide spectrum of human pathophysiological conditions, including infection, aging, cancer, and neurodegenerative disease. There are two major systems in eukaryotic cells that degrade cellular components: the ubiquitin proteasome system (UPS) and the lysosome. Specifically, the UPS degrades proteins, mainly short-lived proteins, that must be tagged by ubiquitin to be recognized by the proteasome [2]. The lysosomal system is responsible for degrading macromolecules, including proteins, and the turnover of organelles by autophagy [3]. Autophagy is an evolutionarily conserved, catabolic quality control process that maintains cellular homeostasis by degrading damaged organelles, misfiled protein aggregates, and foreign organisms. Specifically, autophagy maintains cellular protein homeostasis by facilitating the removal of ubiquitinated protein aggregates. It serves as an overflow pathway for protein turnover under impaired proteasomal activity conditions [4]. Therefore, autophagy seems closely to be associated with oral mucosal homeostasis dysregulation. In this chapter,

Inflammation and Oral Cancer. https://doi.org/10.1016/B978-0-323-88526-3.00006-3
Copyright © 2022 Elsevier Inc. All rights reserved.

101

we introduce oral homeostasis and autophagy's roles in dysregulated homeostatic conditions, such as oral infection, aging and cellular senescence in the oral epithelium, and oral cancers.

2. Oral mucosal homeostasis

2.1 Oral mucosal architecture

The oral mucosa is the mucous membrane lining the inside of the mouth, resembling skin. The oral epithelium resembles, in great part, the architecture of its epidermal counterpart in the skin. An important difference between the two is that the oral epithelium enhances permeability because of the absence of a prominent granular layer, which serves as a physical barrier in skin [5]. The oral mucosa is surfaced by distinct types of mucosal squamous epithelium, depending on anatomical and functional characteristics in different areas. The three main types of oral mucosa are lining mucosa, lingual mucosa, and masticatory mucosa, which includes the dentogingival mucosa [6]. Masticatory and lingual mucosa are surfaced by a parakeratinized epithelium, which lacks a prominent granular layer and covers regions exposed to strong shear forces such as the tongue, attached gingiva, and hard palate. The lining mucosa contains nonkeratinized epithelia and lines the remainder of the oral cavity.

2.2 Maintenance of oral mucosal homeostasis

Despite high permeability, the oral mucosal epithelium is the first barrier protecting the underlying tissue from microbial invasion, toxins, and mechanical forces. The squamous epithelium has structural properties, like stratification and cornification of keratinocytes and specific cell–cell interactions, to maintain its barrier. Recent studies show that epithelial cells are not passive bystanders; rather, they are metabolically active and capable of reacting to external stimuli by immunologic reaction and synthesis of a number of cytokines, adhesion molecules, growth factors, and chemokines [7, 8]. Recently, in line with intestinal epithelial cells in the gastrointestinal tract, oral mucosal epithelial cells can produce antimicrobial peptides, such as defensins and LL-37, and participate in microbiota regulation [9]. In chronic periodontitis, these antimicrobial peptides may be deregulated, allowing uncontrolled growth of pathogenic bacteria [10, 11]. These active reactions of oral epithelial cells help maintain oral mucosal homeostasis. In fact, oral mucosal disorders are led by the dysregulation of oral mucosal homeostasis, destroying the oral epithelial barrier.

3. Autophagy

3.1 Five steps in autophagosome formation

Autophagy is an evolutionarily conserved recycling pathway found in unicellular organisms, such as yeast, and a broad range of complex multicellular organisms, spanning flies, worms, and mammals [12]. It is a dynamic intercellular process by which cytosolic materials, including damaged organelles, toxic protein aggregates, and pathogens, are sequestered into specialized double membrane-bound autophagosomes, and then delivered to the lysosome for degradation [3, 13]. There are three major subtypes of autophagy in the mammalian system, the most common of which is macroautophagy (hereafter referred to as autophagy), as well as microautophagy and chaperone-mediated autophagy [14]. Autophagosome production is divided into five distinct steps: (1) initiation, (2) autophagosome nucleation, (3) expansion and elongation of the autophagosome membrane, (4) closure and lysosome fusion, and (5) degradation of intravesicular products (Fig. 1) [15]. In the initiation step, autophagy-related 13 (ATG13) anchors Unc-51 like autophagy activating kinase 1 (ULK1) to a preautophagosomal structure (PAS), and then almost all autophagy-related proteins gather onto the PAS, hierarchically. The PAS plays a crucial role during autophagy induction because it is a dock structure for autophagy-related (ATG) protein recruitment [16, 17].

Mature autophagosome formation includes nucleation of multiple ATG proteins at the PAS, followed by isolation membrane elongation and autophagosome maturation, all of which requires four different types of functional units. Multiple proteins of the class III PI3K complex, namely, beclin-1, vacuole protein sorting 34 (VPS34), VPS15, and ATG14L, as well as ATG9A, ATG2, and WIPI1/2 of the ATG9A system, accumulate onto the PAS. This accumulation

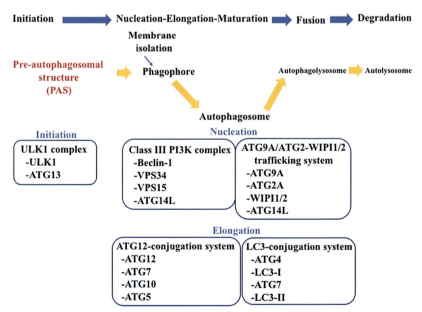

FIG. 1 Schematic overview of the autophagic process. The production of autophagosomes is divided into five steps: initiation, nucleation, elongation, fusion, and degradation.

104 Inflammation and oral cancer

leads to the formation of a phagophore (also known as an isolation membrane) in the nucleation process. Once the first small ATG9A-positive vesicles fuse at the PAS to form a phagophore, the bowl-shaped membrane is elongated continuously, wrapping and engulfing portions of the cytoplasm and organelles. The phagophore's expansion into an autophagosome is generated by ATG5 and ATG12 ubiquitin-like conjugation, which is formed by the activation of ATG7 [18]. Finally, the mediation of two ubiquitin-like ATG conjugation systems, ATG12-ATG5 and ATG8/LC3 [19], causes the isolation membrane to form a closed bilayer membrane structure, resulting in a mature autophagosome with an inner and outer membrane. The ATG5-ATG12-ATG16L1 complex catalyzes the conjugation of microtubule-associated protein 1 light chain 3 (LC3) to phosphatidylethanolamine on the autophagic membrane, forming the autophagosome [20]. This complex also facilitates the lipidation of cytosolic LC3-I into autophagosome membrane-bound LC3-II, a molecular signature of an autophagosome [21]. LC3-II-positive autophagosomes immediately dock and fuse with the lysosome to form an autophagolysosome. Using many lysosomal enzymes, including lysosomal hydrolases, the autophagolysosome degrades the inner membrane of the autophagosome. Cytoplasm-derived macromolecules, such as proteins and organelles, in the autophagosome are broken down into their constituent amino acids or peptides for reuse by the cell.

3.2   Signaling pathways of autophagy

Multiple signaling pathways tightly regulate the autophagy process, which is induced in response to diverse environmental stimuli [12]. The key nutrient sensor mammalian target of rapamycin (mTOR) is an important negative regulator of autophagy [22]. mTOR decreases autophagy activity by inhibiting ULK1, a key upstream kinase that activates autophagy. Nutrient availability conditions and specific growth factors activate mTOR [23]. Nutrient exhaustion results in decreased cellular ATP and activates adenosine monophosphate-activated protein kinase (AMPK), a well-known kinase that initiates a cellular response against nutrient depletion [24]. Activated AMPK induces autophagy by ULK1 phosphorylation in association with mTOR inactivation [25]. ULK1 activation facilitates trafficking of selective ATG proteins to generative sites of phagophores and/or autophagosomes through direct phosphorylation of these ATGs [26, 27].

3.3   Effect of autophagy on cellular differentiation

In addition to being a critical regulator of cellular homeostasis, autophagy is responsible for regulating cellular differentiation and development. Cellular differentiation often requires the degradation of proteins, nucleic acids, and organelles because cells must switch into the final differentiation state or progress to the next developmental stage. Furthermore, cellular remodeling is an energy-demanding process, often associated with nutrient-deficient phases of cellular differentiation and development. Therefore, autophagy is a useful tool for achieving these tasks. During cellular differentiation and development, autophagy plays important roles in the elimination of entire regions of the cytosol and recycling their properties. Recent reports revealed that autophagy-mediated differentiation events in mammalian cells include removing maternal macromolecules during early embryogenesis [28] and clearing mitochondria during erythrocyte [29, 30], lymphocyte [31], and adipocyte [32] differentiation.

Furthermore, recent studies proposed autophagy's crucial roles in adipogenesis and osteogenesis. The pharmacological inhibition and/or genetic knockdown of autophagy-related molecules suppresses adipocyte differentiation [32–34]. The knockdown of two essential autophagy genes, ATG7 and ATG5, in preadipocytes inhibits lipid accumulation and decreases protein levels of adipocyte differentiation factors. These results indicate that an autophagic function drives adipogenesis. In an in vivo experiment, an adipocyte-specific mouse knockout of ATG7 generated lean mice with decreased white adipose mass and increased brown adipocyte features. Furthermore, an increase in the mass of normal brown adipose tissue led to a significantly increased rate of fatty acid β-oxidation. Therefore, because of autophagy loss in adipocytes, these effects translate into a leaner animal with increased insulin sensitivity, suggesting that inhibiting autophagy by disrupting ATG7 has a unique antiobesity and insulin-sensitization effect [34]. These findings suggest that autophagy's function in regulating adipocytes' differentiated state and metabolic function is of a significant magnitude, capable of altering whole-body energy homeostasis.

3.4 Regulation of osteogenesis by autophagy

Increasing evidence indicates autophagy has a role in osteogenesis [35–40]. mTOR has been implicated in controlling mesenchymal stem cell (MSC) fate [41] and osteoblastic differentiation. As a vital intracellular molecule, mTOR plays crucial roles as an extracellular energy sensor, a cell proliferation regulator in a variety of cells and tissues, and a negative autophagy regulator. Studies using either the mTOR inhibitor rapamycin (Rap) or mTOR gene silencing revealed that mTOR signaling exerts both stimulatory [37] and inhibitory [40, 42, 43] effects on osteoblastic differentiation. In the context of mTOR as an osteoblastic differentiation stimulator, Rap treatment markedly reduced mRNA expression levels of osteoblast-specific markers (osteocalcin, bone sialoprotein, and osterix) and decreased RUNX2 (runt-related transcription factor 2) abundance, alkaline phosphatase (ALP) activity, and mineralization capacity of preosteoblastic cells and bone marrow stromal cells. Rap pretreatment had no effect on osteogenic factors in differentiated osteoblasts. These findings suggest that mTOR signaling affects osteoblastic functions by targeting osteoblast proliferation and the early stage of osteoblast differentiation.

In contrast, other studies revealed that Rap treatment promoted osteoblastic differentiation of human embryonic stem cells and bone marrow mesenchymal stem cells [38, 40]. Although mTOR is a member of the PI3K/Akt/mTOR signaling pathway, it has an important role in regulating self-renewal and differentiation in human stem cells. Rap potently inhibits the activities of mTOR and its substrate p70S6K in undifferentiated stem cells. In stem cells treated with Rap for 2–3 weeks, osteoblastic differentiation was further characterized by the expression of osteoblastic marker mRNAs and/or proteins (osterix, osteocalcin, osteoprotegerin, osteonectin, and bone sialoprotein), ALP activity, and alizarin red S staining for mineralized bone nodule formation. These characteristics suggest that Rap is a potent stimulator of osteoblastic differentiation of stem cells and modulates Rap-sensitive mTOR and BMP/Smad signaling [38, 40]. However, these conflicting results on the functions of mTOR in osteogenesis may reflect stage-specific roles of mTOR within the osteoblast lineage.

Moreover, recent studies have addressed AMPK's contribution to osteogenesis. AMPK is a principal intracellular energy sensor that activates energy-producing pathways and inactivates energy-requiring pathways when the cellular AMP/ATP ratio increases [44]. By a new mechanism to control mammalian autophagy, AMPK induces autophagy mainly through phosphorylation of its downstream target raptor and consequent inhibition of mTOR. In turn, mTOR is a known negative autophagy regulator that promotes ULK1 activity, which can initiate the autophagy process [45]. AMP-controlled osteogenic differentiation of human MSCs provided evidence of AMPK-mediated regulation of autophagy through both early mTOR inhibition-mediated autophagy and late activation of the Akt/mTOR signaling axis [39]. Similarly, Yasunaga et al. investigated the role of AMPK in autophagy stages through interplay of the AMPK/mTOR/beclin-1 signaling pathways in osteogenic differentiation of human periodontal ligament stem cells (HPLSCs). They combined pharmacological inhibition and genetic knockdown approaches [35] and concluded that AMPK-dependent mTOR inhibition and beclin-1 activation-mediated early stages of autophagy are required to enhance osteogenic differentiation of HPLSCs. They noted that these elements may be targeted to advance mesenchymal stem cells' osteogenic capacity and provide unique strategies for bone regeneration and repair.

4. Effect of autophagy on oral infection

4.1 Oral infection

The oral cavity is a significant potential source of infection that contributes to the total disease burden and affects overall health and wellbeing. However, in general, we are not seriously concerned about oral infections, as they usually last a few days at most and can be treated easily by simple home remedies. The oral microbiota is both rich and unique. Indeed, a similar microbiota does not exist elsewhere in the body. In healthy persons, *viridans streptococci* comprise most indigenous oral flora. Among all microorganisms in the oral cavity, some periodontal pathogens, such as *Aggregatibacter actinomycetemcomitans*, *Poryphyromonas gingivalis*, *Tannerella forsythia*, and *Treponema denticola*, can induce inflammatory responses that lead to gingivitis and periodontitis [46]. Oral infection spread is generally confined by anatomical barriers or tissue planes, such as muscle and bone. However, infections can spread back and down to the larynx and mediastinum. Particularly in older adults, oral periodontopathic bacteria can cause aspiration pneumonia. Aspiration pneumonia is caused by bacterial clusters from foreign material descending into the branchial tree and lung alveoli. When they originate from the oral cavity, most consist of food debris, saliva, biofilm, or any combination. The airways are protected by coughing and ciliary transport as well as intact immune mechanisms in healthy adults. Age and functional decline impair these defense mechanisms, rendering older persons more vulnerable to developing aspiration pneumonia [47].

4.2 What is xenophagy?

As numerous bacterial pathogens can invade host cells or attempt to break through host defense systems containing mechanical and innate immune barriers, infected cells must have antibacterial tools. In addition to its housekeeping role to maintain cellular homeostasis via nutrient recycling and degrading damaged or aged cytoplasmic constituents, accumulating evidence shows autophagy also functions as an intracellular innate defense mechanism, responding to various bacterial and viral infections. The autophagic degradation of pathogens has been termed xenophagy [48]. Usually, internalized bacteria first localize within a vacuolar compartment to drive the infection process. However, some pathogens, such as *Listeria* and *Shigella*, have evolved mechanisms to escape from their internalization vacuole to the cytosol and avoid destruction by phagolysosomes. Furthermore, other pathogens, like *Salmonella* and mycobacteria, interfere with the normal biogenesis of phagolysosomes to form replicative vacuoles. Xenophagy has been highlighted as a fundamental host cell response to bacterial invasion in vitro by degrading intracellular pathogens located both in the cytosol and within an internalization vacuole, including *Listeria monocytogenes* [49], *Shigella flexneri* [50], *Salmonella typhimurium* [51], and *Mycobacterium tuberculosis* [52]. These studies showed that the mechanisms cells use to target intracellular bacteria to autophagosomal compartments are identical to those used for selective autophagy of endogenous cargo. Cellular cargo is commonly targeted to autophagosomes by interactions between a molecular tag, such as polyubiquitin, and adaptor proteins, such as p62 (also known as SQSTM1), NBR1, and LC3. These adaptor molecules enable xenophagy to selectively target designated cargo to LC3-positive isolation membranes. A similar mechanism involving ubiquitin and p62 appears to be involved in the targeting of intracellular bacteria.

4.3 Xenophagy in periodontal infection

In oral infection, a significant number of periodontitis-related bacteria, such as Gram-negative *Bacteroides forsythus* and *Porphyromonas gingivalis* (Pg), are found in dental plaque in the gingival sulcus [53]. Periodontitis is the most prevalent inflammatory condition among oral diseases. It is typically characterized by the breakdown of tooth-supporting tissues, resulting in the loss of dentition. [54]. Recently, Hagio-Izaki et al. [55] demonstrated xenophagy's role in epithelial keratinocytes (KCs) of the gingival sulcus. It is directly affected by periodontal bacteria, leading to the adhesion and invasion of bacteria into sulcular epithelial KCs. Autophagy in both exfoliative sulcular and cultured KCs is induced by lipopolysaccharide (LPS), the biologically active constituent of endotoxins derived from the cell wall of Gram-negative bacteria (Fig. 2). Exfoliative epithelial KCs showed LPS-induced protein expression and bacteria colonization with autophagosomes. In cultured KCs, Pg-originated LPS (PgLPS)-induced autophagy was related to upregulation of AMPK activity through reactive oxygen species (ROS) and associated with toll-like receptor (TLR)-4-dependent signaling. Moreover, PgLPS enhanced internalization of bacteria-loaded bioparticles and promoted colocalization of the particles with autophagosomes. In contrast, treatment with the autophagy inhibitor 3-methyladenine (3MA) significantly attenuated internalization of bacterial-loaded bioparticles and their colocalization with autophagosomes induced by PgLPS treatment. These findings suggest that PgLPS-induced xenophagy is at least partially responsible for the interaction between periodontal bacteria and

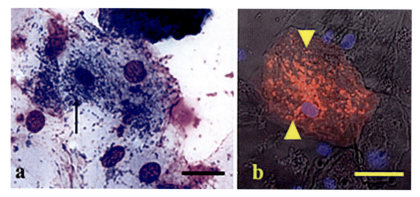

FIG. 2 Intracellular detection of bacteria and autophagosomes in exfoliative sulcular keratinocytes. (A) Representative images of exfoliative epithelial keratinocytes stained with May-Giemsa staining. The *arrow* indicates intracytoplasmic bacterial particles. Scale bar=20μm. (B) Immunocytochemical evaluation of LC3 expression in exfoliative sulcular keratinocytes. LC3-positive autophagosomes are shown as *red-colored particles* in the cytoplasm of sulcular cells. The nuclei were stained with Hoechst 33342 (*blue*). Arrowheads indicate autophagosomes. Scale bar=20μm. *Adapted from Hagio-Izaki K, Yasunaga M, Yamaguchi M, et al: Lipopolysaccharide induces bacterial autophagy in epithelial keratinocytes of the gingival sulcus. BMC Cell Biol 2018;19(1):18.*

sulcular KCs in the gingival sulcus. Despite convincing evidence accumulated in vitro, xenophagy's role in the control of pathogens in vivo remains unclear.

5. Effect of autophagy on aging and cellular senescence

5.1 Aging and autophagy

Aging is characterized by a progressive loss of physiological integrity, which is defined as a time-dependent functional decline, leading to impaired cell and tissue function and increased vulnerability to death. The hallmarks of aging include genetic instability, telomere attrition, epigenetic alterations, loss of proteostasis, deregulated nutrient sensing, mitochondrial dysfunction, cellular senescence, stem cell exhaustion, and altered intercellular communication (Fig. 3) [56]. These factors cause a progressive drop in body quality by homeostatic dysregulation. Previous studies suggest that aging may be controlled by a genetic-hormone system [56, 57]. The insulin and insulin growth factor-1 (IFG-1) signaling (IIS) pathway plays a major role in longevity. IIS pathway attenuation by different genetic manipulations consistently extends the life span of worms, flies, and mice [58]. Among the downstream effectors of the IIS pathway, the most relevant one for longevity was the transcription factor FOXO

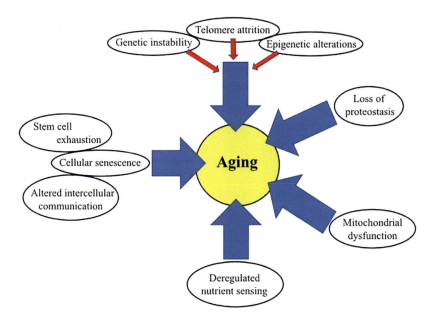

FIG. 3 The hallmarks of aging. Schema showing the conceivable hallmarks of aging, including genetic instability, telomere attrition, epigenetic alterations, loss of proteostasis, mitochondrial dysfunction, deregulated nutrient sensing, stem cell exhaustion, cellular senescence, and altered intercellular communication. *Adapted from Lopez-Otin C, Blasco MA, Partridge L, et al: The hallmarks of aging. Cell 2013;153(6):1194–1217.*

110 Inflammation and oral cancer

[59, 60]. There are four FOXO protein members in mice, but the effect of their overexpression on longevity, and their role in mediating increased health span through reduced IIS, are unclear.

In addition to the IIS pathway, which participates in glucose sensing, there are three additional related and interconnected nutrient sensing systems, containing mTOR, AMPK, and sirtuin 1 (SIRT1) [61]. The kinase mTOR detects high amino acid concentrations and essentially regulates all aspects of anabolic metabolism [62]. Furthermore, the genetic downregulation of mTOR extends longevity in yeast, worms, and flies [63]. In mice, treatment with the mTOR inhibitor Rap also extends longevity in a significantly robust chemical intervention to increase lifespan in mammals [64]. Therefore, mTOR downregulation appears to be the critical mediator of longevity in relation to mTOR. AMPK and SIRT1 act antagonistically to IIS and mTOR, indicating their role in signaling nutrient scarcity and catabolism instead of nutrient abundance and anabolism. In addition, their upregulation favors healthy aging.

5.2 Cellular senescence and autophagy

Cellular senescence is a permanent phenomenon of cell cycle arrest that helps remodel tissue during development and after injury, and is involved in tumor suppression, biological aging, and development of proinflammatory age-related diseases [65]. Cellular senescence is characterized by an altered cell state that is activated in response to persistent DNA damage triggers by various stimuli, such as telomere shortening, activated oncogenes, oxidative stress, and cell–cell fusion. There are two types of cellular senescence: replicative senescence and premature senescence. Replicative senescence is triggered by DNA damage response (DDR) due to telomere shortening. Premature senescence is an active cytostatic program that is triggered in response to types of proliferative or genotoxic stress, such as strong oncogene expression, tumor suppressor loss, exposure to DNA damage, and reactivation of tumor suppressor pathways [65, 66]. Although cellular senescence evolved to eliminate damaged cells, the full sequence of senescence-clearance-regeneration is not entirely accomplished. As a result, senescence may become part of the problem rather than its solution [67].

In vitro, senescent cells show general morphological changes, becoming large, flat, vacuolized, and occasionally, multinucleate. However, in vivo, senescent cells maintain the normal morphology dictated by tissue architecture. Several biomarkers are generally combined to define senescence in both cultured cells and tissues [68]. The most widely used assay for senescent cells is the cytochemical and/or histochemical detection of β-galactosidase activity at pH 6.0, known as senescence-associated β-galactosidase (SAβGal) [69]. This activity is based on the increased lysosomal content of senescent cells, enabling the detection of lysosomal βGal at a suboptimal pH (pH 6.0), and likely reflecting the increased autophagy performed by senescent cells by enlarging the lysosomal compartment [70]. Although we cannot conclude that cells are senescent based on SAβGal activity alone, the assay is usually combined with staining for additional markers such as γH2Ax, a known marker for DDR activation. Similarly, another study reported senescent cells being positively stained with Sudan Black B, which detects the complex lysosomal aggregate known as lipofuscin [71]. In addition, negative markers that should be absent in senescent cells can be used to exclude the cells that are not senescent. These markers indicate cell proliferation, like Ki67 or BrdU incorporation, or proteins like HGMB1. The markers are ubiquitously present in cell nuclei, but secreted from senescent cells, and thus, are absent in the nucleus [72]. Other canonical senescence markers comprise the most common cellular senescence mediators, including p16, ARF/p53, p21, p15, p27, and hyperphosphorylated RB. Heterochromatin foci are also a feature of some senescent cells and are known as senescence-associated heterochromatin foci [73].

Recent studies introduced an intriguingly close relationship between autophagy and its related factors and cellular senescence. Autophagy and senescence share several characteristics, suggesting that both responses may collaterally protect the cell from the toxicity of external stresses, such as radiation and chemotherapy, and from internal forms of stress, such as telomere shortening and oncogene activation. However, it remains unclear whether autophagy acts as a positive or negative regulator of cellular senescence. In support of autophagy's antisenescence role, autophagy suppression induced by knockdown of ATG7 and ATG5 conferred cellular senescence in human fibroblasts due to mitochondrial impairment and the accumulation of produced ROS [74]. Similarly, unsuccessful autophagic activity promotes cellular senescence, as defined by these cells' numerical and functional decreases [75]. The upregulation or downregulation of autophagy through Rap or 3MA treatment, respectively, was employed to explore premodulated autophagy on cellular senescence before inducing D-galactose-mediated cellular senescence [76]. The use of Rap for 24 h significantly reduced cellular senescence, and this change was accompanied by decreased ROS production. Furthermore, p-Jun N-terminal kinase and p38 expression was suppressed in Rap-treated cells [76]. In addition, Rap's antisenescence role was counteracted by the increasing ROS level. Using p38 inhibitors reverted to the oxidative stress-induced senescence of H_2O_2 on cells.

Conversely, autophagy markers do occur in senescent cells. Previous studies revealed a pronounced impact of autophagy inhibition through silencing of ATG7 and an autophagy blockade using 3MA suppressed the senescent population [77, 78]. Furthermore, another study found that knockdown of ATG5 in MCF-7 and U2OS cells treated with curcumin inhibited autophagy by delaying cellular senescence [79]. Maddiodi et al. [80] reported on the BRAF oncogene's capacity to promote both autophagy and senescence in melanoma cells, where autophagy was associated with mTOR inhibition. Similarly, a recent study by Yamaguchi et al. [81] using human keratinocytes treated with H_2O_2, also reported that oxidative stress induced both autophagy and cellular senescence in those cells. Furthermore, attenuation of LC3-II expression using 3MA suppressed oxidative-induced autophagy and senescence.

The promotion of both autophagy and senescence depends on activation of the p38 mitogen-activated protein kinase α (MAPKα) pathway, mediated by intracellular ROS production. Although there is evidence for a relationship between autophagy and cellular senescence, elucidation of autophagy's exact role in inducing cellular senescence remains unclear. Cells exposed to a variety of stress signals that lead to DDR either undergo apoptosis or enter premature senescence. Interestingly, recent studies postulated that stress-induced autophagy plays a cytoprotective role for DDR-affected cells, since entering cellular senescence prevents apoptosis. Several facts support this possible role, including the fact that autophagy activity is enhanced in senescent cells compared to young cells, autophagy suppression increases apoptosis and hinders the senescence process in the course of senescent induction, and a cytoprotective role of autophagy is consistent with its effect on lifespan extension [66, 78, 82]. Similarly, a study by Guo et al. [66] using fibroblasts exposed to H_2O_2 or RasV12 demonstrated that autophagy inhibition promotes apoptosis in senescent cells, suggesting that autophagy activation plays a cytoprotective role. Furthermore, they linked increased autophagy in senescent cells to activation of the transcription factor Foxo3A, which blocks ATP generation by transcriptionally upregulating the expression of PDK4, leading to AMPK activation and mTOR inhibition. Furthermore, Slobodnyuk et al. [82] recently supported the hypothesis that stress-induced autophagy via p38α activity favors cellular senescence over apoptosis (Fig. 4). Their findings indicated that the level of p38α-induced autophagy may determine whether cells undergo apoptosis or senescence in response to stress.

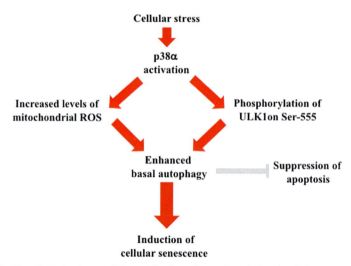

FIG. 4 Hypothetical schema of the mechanism of autophagy-induced cellular senescence by p38α. The p38α pathway activates under cellular stress conditions, increasing mitochondrial ROS levels and ULK1 (Ser555) phosphorylation. These events enhance basal autophagy, inducing cellular senescence instead of apoptosis.

6. Autophagy in cancer

Autophagy modulation is a hallmark of cancer cells. There is abundant evidence that autophagy plays a role in each tumorigenesis stage. Depending on the type of cancer and the context, autophagy attributes a dual role, having both tumor-suppressing and tumor-promoting properties (Fig. 5). Autophagy deregulation, which leads to tumor suppression, is prevalent in many types of cancer and involves several autophagy-related genes and proteins [83, 84]. Paradoxically, autophagy's cytoprotective role may be detrimental to the treatment of cancer cells since it helps them resist anticancer therapy [85, 86].

6.1 Effects of autophagy on tumor metabolism

The Warburg hypothesis was that cancer growth is caused by the fact that tumor cells mainly generate energy in the form of ATP by the nonoxidative breakdown of glucose. This contrasts with the understanding that normal cells produce ATP during oxidative phosphorylation, thereby obtaining fuel through the oxidative breakdown of glucose [87]. Thus, the Warburg effect provides that enhanced conversion of glucose to lactate occurs in tumor cells, even in the presence of normal levels of oxygen. The PI3K pathway is considered the master regulator of the glycolytic phenotype through AKT1 and mTOR signaling and HIF-1 activation. The AMPK pathway is also important because AMPK is considered a metabolic checkpoint. It controls cell proliferation when activated under energetic stress. AMPK activation enhances

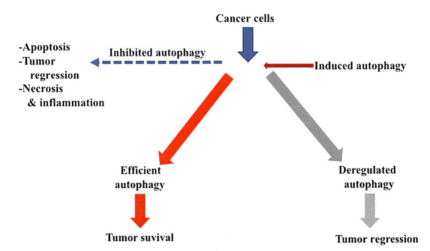

FIG. 5 Effects of autophagy in cancer development. Autophagy may be associated with a dual role in cancer cells. Efficient autophagy conditions enable cancer cells to survive and proliferate, whereas deregulated autophagy leads to tumor suppression and apoptosis. Furthermore, when autophagy is inhibited, the activation of apoptosis or necrosis occurs in cancer cells.

autophagy to inhibit mTOR activity and accelerates both ULK1 and beclin-1 activity. Furthermore, a deep metabolic reprograming of oncogenic transformations helps upregulate aerobic glycolysis in favor of oxidative phosphorylation [88]. Human cancer cell lines bearing activating mutations in RAS commonly have high levels of basal autophagy. Suppressing essential autophagy proteins, such as ATG5 and ATG7, abolishes tumorigenicity and reduces glucose metabolism, indicating their high dependence on autophagy for survival [89, 90]. Therefore, autophagy may contribute to metabolism rearrangement by targeting the degradation of damaged mitochondria, thereby decreasing oxidative phosphorylation.

6.2 Cancer stem cells (CSCs)

Similar to normal stem cells, CSCs can renew themselves and have a limited differentiation capacity [91]. Since CSC theory's establishment and the discovery of CSCs in individual cancer types, autophagy has been proposed as a key mechanism in their homeostasis, dismissal, and spread. Comparing breast CSCs to adherent cells indicated that cancer sphere assay upregulated autophagy, making it a useful tool for CSC detection [92, 93]. Therefore, two key autophagy proteins, beclin-1 and ATG4, were both required for the maintenance and expansion of CSCs. Recently, a number of studies demonstrated that autophagy impairment negatively affects the expression of stamina markers, and consequently, the self-renewal capacity of cells in a variety of CSCs, such as breast [94], pancreatic, liver [95], osteosarcoma [96], ovarian [97], and glioblastoma [98]. In chronic myelogenous leukemia, some autophagy-related genes, such as ATG4, ATG5, and beclin-1 [99–101], are upregulated. Silencing both ATG7 and ATG4B affects cell survival, suggesting that autophagy acts in both chemoresistance and tumor-suppressive mechanisms. To better understand the molecular mechanisms of autophagy-dependent CSC maintenance, Yeo et al. [102] demonstrated that autophagy acts through the EGFR/STAT3 and Tgfβ/Smad signaling pathways in the distinct breast cancer stem-like cells, ALDH$^+$ and CD29hiCD61$^+$, respectively. By FIP200 depletion, they showed decreased EGFR phosphorylation, which decreased STAT3 activation, and consequently, impaired ALDH$^+$ cell tumorigenicity. In contrast, autophagy inhibition decreased expression of TFGβ2 and TGFβ3 and induced a defect in Smad signaling, which is indispensable for the CD29hiCD61$^+$ phenotype in cells. Recently, Sharif et al. [103] found that any perturbation in basal autophagy, generated using either autophagy inhibitors or activators, suppresses tetracarcinoma CSCs' pluripotency, leading to differentiation and/or senescence. In ovarian CSCs, there is a crucial link between autophagy and stemness [97]. Interestingly, some studies suggested autophagy's role in regulating chromosome stability by coordinating ATR checkpoint and double-strand-break processing. Thus, CSCs may exploit autophagy to prevent further DNA damage after an initial insult and maintain their survival [104].

6.3 Tumor-suppressing function of autophagy

Tumor suppressor genes, such as PTEN, AMPK, LKB1, and TSC1/2, negatively regulate mTOR, which participates in the regulation of cell growth and stimulates autophagy. Conversely, oncogenes, such as class I PI3K, Was, RHEB, and AKT, activate and inhibit mTOR [105]. As a member of the class III PI3K complex that enhances autophagy, beclin-1 is a tumor suppressor in mammalian cells. Monoallelic mutations in beclin-1 are frequently observed in human breast, ovarian, and prostate cancers. Studies in mice have demonstrated that mice are more sensitive to spontaneous tumor development when beclin-1 is monoallelically disrupted. These findings provide direct evidence for beclin-1's role as a haploinsufficient tumor suppressor gene that is implicated in the pathogenesis of several human cancers [106, 107]. Additionally, death-associated protein kinase, which disrupts the beclin-1/BCL-2 complex by induction of beclin-1 phosphorylation, is commonly silenced in different types of human cancers by methylation [108]. BCL-2 and BCL-XL are antiapoptotic members of the BCL-2 family that modulate cell death in an autophagy-independent manner. They are overexpressed in several hematological malignancies [109]. They suppress cell death and promote survival and growth of cancer cells by suppressing BAK/BAX-dependent pore formation during mitochondrial outer membrane permeabilization [110]. In addition to BCL-2 and BCL-XL's roles in apoptosis inhibition, they have been implicated in oncogenesis as negative autophagy regulators, negatively participating in mTOR signaling and directly interacting with the BH3 domain of beclin-1 [111].

The first evidence indicating a link between autophagy and tumor suppression came from beclin-1 studies. Monoallelic loss of beclin-1 on chromosome 17q21 was reported in 40%–75% of human breast, ovary, and prostate tumors, suggesting that autophagy has a tumor suppressor role [106, 107]. Furthermore, downregulation of beclin-1 protein levels occurred in various types of brain cancer [112]; and in beclin-1$^{+/-}$ mice, there was a high incidence of spontaneous tumors, especially lymphoma and hepatocellular carcinoma. However, it remains controversial whether beclin-1 is a tumor suppressor gene. Laddha et al. [113] suggest that, contrary to these previous reports, beclin-1 is not significantly mutated in human cancer; therefore, it is not a classic tumor suppressor gene. Under specific conditions, increased autophagic flux causes autophagic cell death and explains, at least in part, autophagy's tumor suppressor effect [114]. Pattingre et al. [111] reported that the expression of a mutant beclin-1, which was unable to interact with BLC-2, induced autophagy to a greater extent than wild-type beclin-1 and promoted cell death. Recently, studies using an ovarian cancer cell line showed that ectopic RAS expression induces autophagic cell death by upregulating beclin-1. Similarly, Zhao and colleagues demonstrated that the transcription factor Foxo1 facilitates autophagy to induce autophagic cell death in tumor cells. These results indicate that autophagy promoted by cytosolic Foxo1 suppresses tumors [115, 116]. Research on knockout mice for genes encoding

ATG protein family members associated defects in specific regulators of this process with the development of a tumorigenic phenotype. Mice with systemic mosaic Atg5 deletion or liver-specific deletion of Atg7 spontaneously develop benign liver adenomas [117]. These findings suggest that autophagy defects promote the development of benign tumors in this tissue. However, they also indicate that the absence of autophagy prevents progression to a malignant phenotype. Another study revealed that the deletion of Atg7 in mice expressing an activating mutation of BRAF promotes early tumor development in the lung and inhibits progression to a more malignant phenotype, increasing mouse survival [105]. These findings support the hypothesis that autophagy plays an important role in tumor suppression at early stages. However, the findings discussed also reveal the potentially dual nature of this process in tumor development and progression.

6.4 Tumor-promoting function of autophagy

Under metabolic stress conditions, autophagy is essential for tumor cell survival. Direct or indirect genetic autophagy inactivation suppresses the survival of nutrient-deprived tumors, even when apoptosis is inactivated. Direct phenomena include the allelic loss of beclin-1 as well as RNA interference or deficiency in autophagy-related genes, whereas an indirect means is constitutive activation of the PI3K pathway [89]. The tumor cells need a mechanism to adapt to ischemia in an environment that is both hypoxic and deprived of growth factors and nutrients. In a hypoxic microenvironment, autophagy provides an alternative source of energy for tumor cell survival. Under hypoxic conditions, cells express hypoxia-inducible factor-1 alpha (HIF-1α) [118], which upregulates several genes and facilitates AMPK activation, leading to HIF-1-independent hypoxia-induced autophagy [119]. Therefore, hypoxia-induced autophagy in malignant cells confers both resistance to hypoxia stress and increased tumorigenicity [120]. ROS mediates starvation-induced autophagy [121, 122]. The AMPK-dependent pathway helps modulate starvation-induced autophagy. AMPK activation inhibits the mTOR pathway, initiating autophagy [123]. These findings indicate that starvation-induced autophagy acts as a backup energy reserve in case of starvation.

6.5 Role of autophagy in tumor cell dissemination and metastasis

Autophagy may also promote tumor cell dissemination and metastasis by protecting these cells from anoikis, an apoptotic cell death mechanism. The detachment of anchorage-dependent cells from the surrounding extracellular matrix (ECM) induces anoikis. This detachment from the ECM is a critical step in dissemination. Cancer cells activate autophagy during metastatic progression as a mechanism to resist anoikis [124]. Metastasis induces prosurvival autophagy in epithelial cells to protect them from anoikis. ATG depletion enhances cell death and reduces clonogenic recovery after anoikis [125, 126]. Moreover, autophagy may act as a regulator of migration and cellular invasion. Autophagy suppression is associated with reduced invasion and a partial reversion of epithelial to mesenchymal transition (EMT) [127]. From the dual functions of autophagy in cancer, autophagy plays a role in decreasing the chance of metastasis because it prevents p62-dependent stabilization of the EMT-promoting transcription factor twist-1 [128]. Nonmetastatic cells reduce autophagy activity in association with EMT promotion [120].

6.6 Autophagy as a therapeutic tool for cancer

Autophagy tightly regulates the formation of autophagosomes, which is divided into five distinct stages: (1) initiation, (2) autophagosome nucleation, (3) expansion and elongation of the autophagosome membrane, (4) closure and fusion with the lysosome, and (5) degradation of intravesicular products. Parsing these different formation stages for autophagosomes is important. Each stage has a subset of potential therapeutic targets for inhibiting autophagy in humans.

6.7 Therapeutic targets at the early stages of autophagy

SBI 0206965 (SBI) [129] is one of various ATP-competitive inhibitors against the kinase ULK1 [130]. It inhibits autophagy and synergies with mTOR inhibition [129]. SBI also effectively facilitates a cytotoxic apoptosis response in lung cancer during nutrient starvation. SB02024 is a VPS34 inhibitor that successfully blocked autophagy [131]. Both ULK1 and VPS34 inhibitors effectively suppress autophagy-dependent CNS tumor cells [132]. The ATG4B inhibitors UAMC-2526 [133] and S130 [134] are also effective at blocking autophagy, leading to the identification of their anticancer effects in xenograft colorectal tumors. Although there is preliminary data on the potential therapeutic effects of these early-stage inhibitors, the data stem largely from early preclinical studies. Therefore, additional research is required.

6.8 Therapeutic targets at the late stages of autophagy

Chloroquine (CQ) and hydroxychloroquine (HCQ) inhibit autophagy by decreasing autophagosome/lysosome fusion [135]. CQ was initially developed and pioneered for malaria treatment [136], whereas HCQ was developed by adding a hydroxyl group to CQ. This modification decreased potential toxicities while maintaining overall effectiveness [137]. The first evidence showing CQ's potential anticancer effect was its toxic effect against lymphoma and melanoma cells [138]. The first clinical trial of CQ in the treatment of patients with glioblastoma revealed significantly improved clinical outcomes for patients [139]. Although clinical trials have started to evaluate CQ and HCQ combined with existing chemotherapeutic approaches in cancer treatment, the results obtained to date have been disappointing. Treatment failure in these trials may be caused by a reduced autophagy-induced ATP release. Also, CQ and HCQ inhibit lysosomal acidification [140]. Recently, the bisaminoquinoline Lys05 and dimeric quinacrine DQ661 were developed as more potent and selective next-generation lysosomal-targeted inhibitors to effect mTOR inhibition [141]. Lys05, which is approximately tenfold more potent than CQ, limited melanoma and colorectal adenocarcinoma growth as a single agent in mouse models [142]. Similarly, DQ661 exhibited impressive in vivo single-agent activity in melanoma and colorectal cancer and in vivo efficacy combined with gemcitabine in pancreatic ductal adenocarcinoma [141].

7. Conclusion

Autophagy has important functions in maintaining and dysregulating oral mucosal homeostasis, dealing with organelle maintenance, protein turnover, and cellular stress response. Although autophagy is generally considered a survival or adaptive process, its relationship to cell death pathways needs clarification. Indeed, autophagy's dual nature may be associated with induction of dysregulated homeostatic conditions in the oral mucosa. A greater understanding of autophagy's relative role in the pathogenesis of specific conditions caused by dysregulated homeostasis is critical before targeting autophagic pathway elements for therapy. Further elucidation of the molecular mechanisms that regulate autophagy may guide the search for and identification of new targets for therapeutic manipulation.

Acknowledgments

We would like to acknowledge Enago (www.enago.jp) for the English language review.

Conflict of interest

The authors declare no conflicts of interest.

References

[1] Moutsopoulos NM, Konkel JE. Tissue-specific immunity at the oral mucosal barrier. Trends Immunol 2018;39(4):276–87.

[2] Ciechanover A, Orian A, Schwartz AL. Ubiquitin-mediated proteolysis: biological regulation via destruction. Bioessays 2000;22(5):442–51.

[3] Mizushima N, Levine B, Cuervo AM, Klionsky DJ. Autophagy fights disease through cellular self-digestion. Nature 2008;451(7182):1069–75.

[4] Lamark T, Johansen T. Aggrephagy: selective disposal of protein aggregates by macroautophagy. Int J Cell Biol 2012;2012:736905.

[5] Allam JP, Stojanovski G, Friedrichs N, Peng W, Bieber T, Wenzel J, et al. Distribution of Langerhans cells and mast cells within the human oral mucosa: new application sites of allergens in sublingual immunotherapy? Allergy 2008;63(6):720–7.

Role of autophagy in dysregulation **Chapter | 6 119**

[6] Novak N, Haberstok J, Bieber T, Allam JP. The immune privilege of the oral mucosa. Trends Mol Med 2008;14(5):191–8.

[7] Groeger S, Meyle J. Oral mucosal epithelial cells. Front Immunol 2019;10:208.

[8] Groeger SE, Meyle J. Epithelial barrier and oral bacterial infection. Periodontol 2000 2015;69(1):46–67.

[9] da Silva BR, de Feitas VA, Nascimento-Neto LG, Carneiro VA, Arruda FV, de Aguiar AS, et al. Antimicrobial peptide control of pathogenic microorganisms of the oral cavity: a review of the literature. Peptides 2012;36(2):315–21.

[10] Li X, Duan D, Yang J, Wang P, Han B, Zhao L, et al. The expression of human beta-defensins (hBD-1, hBD-2, hBD-3, hBD-4) in gingival epithelia. Arch Oral Biol 2016;66:15–21.

[11] Dommisch H, Acil Y, Dunsche A, Winter J, Jepsen S. Differential gene expression of human beta-defensins (hBD-1, −2, −3) in inflammatory gingival diseases. Oral Microbiol Immunol 2005;20(3):186–90.

[12] He C, Klionsky DJ. Regulation mechanisms and signaling pathways of autophagy. Annu Rev Genet 2009;43:67–93.

[13] Cuervo AM, Bergamini E, Brunk UT, Droge W, Ffrench M, Terman A. Autophagy and aging: the importance of maintaining "clean" cells. Autophagy 2005;1(3):131–40.

[14] Ravikumar B, Sarkar S, Davies JE, Futter M, Garcia-Arencibia M, Green-Thompson ZW, et al. Regulation of mammalian autophagy in physiology and pathophysiology. Physiol Rev 2010;90(4):1383–435.

[15] Mizushima N. Autophagy: process and function. Genes Dev 2007;21(22):2861–73.

[16] Kotani T, Kirisako H, Koizumi M, Ohsumi Y, Nakatogawa H. The Atg2-Atg18 complex tethers pre-autophagosomal membranes to the endoplasmic reticulum for autophagosome formation. Proc Natl Acad Sci U S A 2018;115(41):10363–8.

[17] Suzuki K, Kirisako T, Kamada Y, Mizushima N, Noda T, Ohsumi Y. The pre-autophagosomal structure organized by concerted functions of APG genes is essential for autophagosome formation. EMBO J 2001;20(21):5971–81.

[18] Geng J, Klionsky DJ. The Atg8 and Atg12 ubiquitin-like conjugation systems in macroautophagy. 'Protein modifications: beyond the usual suspects' review series. EMBO Rep 2008;9(9):859–64.

[19] Abhyankar S, Gilliland DG, Ferrara JL. Interleukin-1 is a critical effector molecule during cytokine dysregulation in graft versus host disease to minor histocompatibility antigens. Transplantation 1993;56(6):1518–23.

[20] Hanada T, Noda NN, Satomi Y, Ichimura Y, Fujioka Y, Takao T, et al. The Atg12-Atg5 conjugate has a novel E3-like activity for protein lipidation in autophagy. J Biol Chem 2007;282(52):37298–302.

[21] Tanida I, Ueno T, Kominami E. LC3 conjugation system in mammalian autophagy. Int J Biochem Cell Biol 2004;36(12):2503–18.

[22] Ravikumar B, Vacher C, Berger Z, Davies JE, Luo S, Oroz LG, et al. Inhibition of mTOR induces autophagy and reduces toxicity of polyglutamine expansions in fly and mouse models of Huntington disease. Nat Genet 2004;36(6):585–95.

[23] Kim J, Kundu M, Viollet B, Guan KL. AMPK and mTOR regulate autophagy through direct phosphorylation of Ulk1. Nat Cell Biol 2011;13(2):132–41.

[24] Hardie DG. AMPK and autophagy get connected. EMBO J 2011;30(4):634–5.

[25] Egan DF, Shackelford DB, Mihaylova MM, Gelino S, Kohnz RA, Mair W, et al. Phosphorylation of ULK1 (hATG1) by AMP-activated protein kinase connects energy sensing to mitophagy. Science 2011;331(6016):456–61.

[26] Papinski D, Kraft C. Atg1 kinase organizes autophagosome formation by phosphorylating Atg9. Autophagy 2014;10(7):1338–40.

120 Inflammation and oral cancer

[27] Young AR, Chan EY, Hu XW, Kochi R, Crawshaw SG, High S, et al. Starvation and ULK1-dependent cycling of mammalian Atg9 between the TGN and endosomes. J Cell Sci 2006;119(18):3888–900.

[28] Tsukamoto S, Kuma A, Murakami M, Kishi C, Yamamoto A, Mizushima N. Autophagy is essential for preimplantation development of mouse embryos. Science 2008;321(5885):117–20.

[29] Sandoval H, Thiagarajan P, Dasgupta SK, Schumacher A, Prchal JT, Chen M, et al. Essential role for Nix in autophagic maturation of erythroid cells. Nature 2008;454(7201):232–5.

[30] Mortensen M, Ferguson DJ, Edelmann M, Kessler B, Morten KJ, Komatsu W, et al. Loss of autophagy in erythroid cells leads to defective removal of mitochondria and severe anemia in vivo. Proc Natl Acad Sci U S A 2010;107(2):832–7.

[31] Pua HH, He YW. Mitophagy in the little lymphocytes: an essential role for autophagy in mitochondrial clearance in T lymphocytes. Autophagy 2009;5(5):745–6.

[32] Singh R, Xang Y, Wang Y, Baikati K, Cuervo AM, Luu YK, et al. Autophagy regulates adipose mass and differentiation in mice. J Clin Invest 2009;119(11):3329–39.

[33] Tao Z, Liu L, Zheng LD, Cheng Z. Autophagy in adipocyte differentiation. Methods Mol Biol 2019;1854:45–53.

[34] Zhang Y, Goldman S, Baerga R, Zhao Y, Komatsu M, Jin S. Adipose-specific deletion of autophagy-related gene 7 (atg7) in mice reveals a role in adipogenesis. Proc Natl Acad Sci U S A 2009;106(47):19860–5.

[35] Yasunaga M, Kajiya H, Toshimitsu T, Nakashima H, Tamaoki S, Ishikawa H, et al. The early sutophagic pathway contributes to osteogenic differentiation of human periodontal ligament stem cells. J Hard Tissue Biol 2019;28(1):63–70.

[36] Chen J, Long F. mTORC1 signaling promotes osteoblast differentiation from preosteoblasts. PLoS One 2015;10(6), e0130627.

[37] Singha UK, Jiang Y, Yu S, Luo M, Lu Y, Zhang J, et al. Rapamycin inhibits osteoblast proliferation and differentiation in MC3T3-E1 cells and primary mouse bone marrow stromal cells. J Cell Biochem 2008;103(2):434–46.

[38] Wan Y, Zhuo N, Li Y, Zhao W, Jiang D. Autophagy promotes osteogenic differentiation of human bone marrow mesenchymal stem cell derived from osteoporotic vertebrae. Biochem Biophys Res Commun 2017;488(1):46–52.

[39] Pantovic A, Krstic A, Janjetovic K, Kocic J, Harhaji-Trajkovic L, Bugarski D, et al. Coordinated time-dependent modulation of AMPK/Akt/mTOR signaling and autophagy controls osteogenic differentiation of human mesenchymal stem cells. Bone 2013;52(1):524–31.

[40] Lee KW, Yook JY, Son MY, Kim MJ, Koo DB, Han YM, et al. Rapamycin promotes the osteoblastic differentiation of human embryonic stem cells by blocking the mTOR pathway and stimulating the BMP/Smad pathway. Stem Cells Dev 2010;19(4):557–68.

[41] Xiang X, Zhao J, Xu G, Li Y, Zhang W. mTOR and the differentiation of mesenchymal stem cells. Acta Biochim Biophys Sin 2011;43(7):501–10.

[42] Ogawa T, Tokuda M, Tomizawa K, Matsui H, Itano T, Konichi R, et al. Osteoblastic differentiation is enhanced by rapamycin in rat osteoblast-like osteosarcoma (ROS 17/2.8) cells. Biochem Biophys Res Commun 1998;249(1):226–30.

[43] Martin SK, Fitter S, Bong LF, Drew JJ, Gronthos S, Shepherd PR, et al. NVP-BEZ235, a dual pan class I PI3 kinase and mTOR inhibitor, promotes osteogenic differentiation in human mesenchymal stromal cells. J Bone Miner Res 2010;25(10):2126–37.

[44] Hardie DG. AMP-activated/SNF1 protein kinases: conserved guardians of cellular energy. Nat Rev Mol Cell Biol 2007;8(10):774–85.

[45] Schroeder MA, DiPersio JF. Mouse models of graft-versus-host disease: advances and limitations. Dis Model Mech 2011;4(3):318–33.

Role of autophagy in dysregulation **Chapter | 6 121**

[46] Paster BJ, Boches SK, Galvin JL, Ericson RE, Lau CN, Levanos VA, et al. Bacterial diversity in human subgingival plaque. J Bacteriol 2001;183(12):3770–83.

[47] Muller F. Oral hygiene reduces the mortality from aspiration pneumonia in frail elders. J Dent Res 2015;94(3):14S–6S.

[48] Huang J, Brumell JH. Autophagy in immunity against intracellular bacteria. Curr Top Microbiol Immunol 2009;335:189–215.

[49] Py BF, Lipinski MM, Yuan J. Autophagy limits Listeria monocytogenes intracellular growth in the early phase of primary infection. Autophagy 2007;3(2):117–25.

[50] Ogawa M, Mimuro H, Yoshikawa Y, Ashida H, Sasakawa C. Manipulation of autophagy by bacteria for their own benefit. Microbiol Immunol 2011;55(7):459–71.

[51] Birmingham CL, Smith AC, Bakowski MA, Yoshimori T, Brumell JH. Autophagy controls Salmonella infection in response to damage to the Salmonella-containing vacuole. J Biol Chem 2006;281(16):11374–83.

[52] Gutierrez MG, Master SS, Singh SB, Taylor GA, Colombo MI, Deretic V. Autophagy is a defense mechanism inhibiting BCG and mycobacterium tuberculosis survival in infected macrophages. Cell 2004;119(6):753–66.

[53] Griffen AL, Becker MR, Lyons SR, Moeschberger ML, Leys EJ. Prevalence of Porphyromonas gingivalis and periodontal health status. J Clin Microbiol 1998;36(11):3239–42.

[54] Petersen PE, Bourgeois D, Ogawa H, Estupinan-Day S, Ndiaye C. The global burden of oral diseases and risks to oral health. Bull World Health Organ 2005;83(9):661–9.

[55] Hagio-Izaki K, Yasunaga M, Yamaguchi M, Kajiya H, Morita H, Yoneda M, et al. Lipopolysaccharide induces bacterial autophagy in epithelial keratinocytes of the gingival sulcus. BMC Cell Biol 2018;19(1):18.

[56] Lopez-Otin C, Blasco MA, Partridge L, Serrano M, Kroemer G. The hallmarks of aging. Cell 2013;153(6):1194–217.

[57] van Heemst D. Insulin, IGF-1 and longevity. Aging Dis 2010;1(2):147–57.

[58] Fontana L, Partridge L, Longo VD. Extending healthy life span—from yeast to humans. Science 2010;328(5976):321–6.

[59] Slack C, Giannakou ME, Foley A, Goss M, Partridge L. dFOXO-independent effects of reduced insulin-like signaling in Drosophila. Aging Cell 2011;10(5):735–48.

[60] Kenyon C, Chang J, Gensch E, Rudner A, Tabtiang R. A C. elegans mutant that lives twice as long as wild type. Nature 1993;366(6454):461–4.

[61] Houtkooper RH, Williams RW, Auwerx J. Metabolic networks of longevity. Cell 2010;142(1):9–14.

[62] Laplante M, Sabatini DM. mTOR signaling in growth control and disease. Cell 2012;149(2):274–93.

[63] Johnson SC, Rabinovitch PS, Kaeberlein M. mTOR is a key modulator of ageing and age-related disease. Nature 2013;493(7432):338–45.

[64] Harrison DE, Strong R, Sharp ZD, Nelson JE, Astle CM, Flurkey K, et al. Rapamycin fed late in life extends lifespan in genetically heterogeneous mice. Nature 2009;460(7253):392–5.

[65] Biran A, Zada L, Abou Karam P, Vadal E, Routman L, Ovadya Y, et al. Quantitative identification of senescent cells in aging and disease. Aging Cell 2017;16(4):661–71.

[66] Guo L, Xie B, Mao Z. Autophagy in premature senescent cells is activated via AMPK pathway. Int J Mol Sci 2012;13(3):3563–82.

[67] Munoz-Espin D, Serrano M. Cellular senescence: from physiology to pathology. Nat Rev Mol Cell Biol 2014;15(7):482–96.

[68] Collado M, Serrano M. The power and the promise of oncogene-induced senescence markers. Nat Rev Cancer 2006;6(6):472–6.

122 Inflammation and oral cancer

[69] Dimri GP, Lee X, Basile G, Acosta M, Scott G, Roskelley C, et al. A biomarker that identifies senescent human cells in culture and in aging skin in vivo. Proc Natl Acad Sci U S A 1995;92(20):9363–7.

[70] Young AR, Narita M, Ferreira M, Kirschner K, Sadaie M, Darot JF, et al. Autophagy mediates the mitotic senescence transition. Genes Dev 2009;23(7):798–803.

[71] Georgakopoulou EA, Tsimaratou K, Evangelou K, Fernandez Marcos PJ, Zoumpourlis V, Trougakos IP, et al. Specific lipofuscin staining as a novel biomarker to detect replicative and stress-induced senescence. A method applicable in cryo-preserved and archival tissues. Aging (Albany NY) 2013;5(1):37–50.

[72] Davalos AR, Kawahara M, Malhotra GK, Schaum N, Huang J, Ved U, et al. p53-dependent release of Alarmin HMGB1 is a central mediator of senescent phenotypes. J Cell Biol 2013;201(4):613–29.

[73] Narita M, Nunez S, Heard E, Narita M, Lin AW, Hearn SA, et al. Rb-mediated heterochromatin formation and silencing of E2F target genes during cellular senescence. Cell 2003;113(6):703–16.

[74] Kang C, Elledge SJ. How autophagy both activates and inhibits cellular senescence. Autophagy 2015;12(5):898–9.

[75] Garcia-Prat L, Martinez-Vicente M, Perdiguero E, Ortet L, Rodriguez-Ubreva J, Rebollo E, et al. Autophagy maintains stemness by preventing senescence. Nature 2016;529(7584):37–42.

[76] Zhang D, Chen Y, Xu X, Xiang H, Shi Y, Gao Y, et al. Autophagy inhibits the mesenchymal stem cell aging induced by D-galactose through ROS/JNK/p38 signalling. Clin Exp Pharmacol Physiol 2020;47(3):466–77.

[77] Patschan S, Chen J, Polotskaia A, Mendelev N, Cheng J, Patschan D, et al. Lipid mediators of autophagy in stress-induced premature senescence of endothelial cells. Am J Physiol Heart Circ Physiol 2008;294(3):H1119–29.

[78] Gewirtz DA. Autophagy and senescence: a partnership in search of definition. Autophagy 2013;9(5):808–12.

[79] Mosieniak G, Adamoqicz M, Alster O, Jaskowiak H, Szczepankiewicz AA, Wilczynski GM, et al. Curcumin induces permanent growth arrest of human colon cancer cells: link between senescence and autophagy. Mech Ageing Dev 2012;133(6):444–55.

[80] Maddodi N, Huang W, Havighurst T, Kim K, Longley BJ, Setaluri V. Induction of autophagy and inhibition of melanoma growth in vitro and in vivo by hyperactivation of oncogenic BRAF. J Invest Dermatol 2010;130(6):1657–67.

[81] Yamaguchi M, Kajiya H, Egashira R, Yasunaga M, Hagio-Izaki K, Sato A, et al. Oxidative stress-induced interaction between autophagy and cellular senescence in human keratinocytes. J Hard Tissue Biol 2018;27(3):199–208.

[82] Slobodnyuk K, Radic N, Ivanova S, Llado A, Trempolec N, Zorzano A, et al. Autophagy-induced senescence is regulated by p38alpha signaling. Cell Death Dis 2019;10(6):376.

[83] Wong AS, Cheung ZH, Ip NY. Molecular machinery of macroautophagy and its deregulation in diseases. Biochim Biophys Acta 2011;1812(11):1490–7.

[84] Wu WK, Coffelt SB, Cho CH, Wang XJ, Lee CW, Chan FK, et al. The autophagic paradox in cancer therapy. Oncogene 2012;31(8):939–53.

[85] Kimmelman AC. The dynamic nature of autophagy in cancer. Genes Dev 2011;25(19):1999–2010.

[86] Roy S, Debnath J. Autophagy and tumorigenesis. Semin Immunopathol 2010;32(4):383–96.

[87] Warburg O. On the origin of cancer cells. Science 1956;123(3191):309–14.

Role of autophagy in dysregulation Chapter | 6 **123**

[88] Kim JH, Kim HY, Lee YK, Yoon YS, Xu WG, Yoon JK, et al. Involvement of mitophagy in oncogenic K-Ras-induced transformation: overcoming a cellular energy deficit from glucose deficiency. Autophagy 2011;7(10):1187–98.

[89] Mathew R, Karantza-Wadsworth V, White E. Role of autophagy in cancer. Nat Rev Cancer 2007;7(12):961–7.

[90] Hamai A, Codogno P, Mehrpour M. Cancer stem cells and autophagy: facts and perspectives. J Cancer Stem Cell Res 2014;2, e1005.

[91] Lobo NA, Shimono Y, Qian D, Clarke MF. The biology of cancer stem cells. Annu Rev Cell Dev Biol 2007;23:675–99.

[92] Wolf J, Dewi DL, Fredebohm J, Muller-Decker K, Flechtenmacher C, Hoheisei JD, et al. A mammosphere formation RNAi screen reveals that ATG4A promotes a breast cancer stem-like phenotype. Breast Cancer Res 2013;15(6), R109.

[93] Gong C, Song E, Codogno P, Mehrpour M. The roles of BECN1 and autophagy in cancer are context dependent. Autophagy 2012;8(12):1853–5.

[94] Gong C, Bauvy C, Tonelli G, Yue W, Delomenie C, Nicolas V, et al. Beclin 1 and autophagy are required for the tumorigenicity of breast cancer stem-like/progenitor cells. Oncogene 2013;32(18):2261–72.

[95] Song YJ, Zhang SS, Guo XL, Sun K, Han ZP, Li R, et al. Autophagy contributes to the survival of CD133$^+$ liver cancer stem cells in the hypoxic and nutrient-deprived tumor microenvironment. Cancer Lett 2013;339(1):70–81.

[96] Zhang D, Zhao Q, Sun H, Yin L, Wu J, Xu J, et al. Defective autophagy leads to the suppression of stem-like features of CD271(+) osteosarcoma cells. J Biomed Sci 2016;23(1):82.

[97] Peng Q, Qin J, Zhang Y, Cheng X, Wang X, Lu W, et al. Autophagy maintains the stemness of ovarian cancer stem cells by FOXA2. J Exp Clin Cancer Res 2017;36(1):171.

[98] Buccarelli M, Marconi M, Pacioni S, De Pascalis I, D'Alessandris QG, Martini M, et al. Inhibition of autophagy increases susceptibility of glioblastoma stem cells to temozolomide by igniting ferroptosis. Cell Death Dis 2018;9(8):841.

[99] Rothe K, Lin H, Lin KB, Leung A, Wang HM, Malekesmaeili M, et al. The core autophagy protein ATG4B is a potential biomarker and therapeutic target in CML stem/progenitor cells. Blood 2013;123(23):3622–34.

[100] Karvela M, Baquero P, Kuntz EM, Mukhopadhyay A, Mitchell R, Allan EK, et al. ATG7 regulates energy metabolism, differentiation and survival of Philadelphia-chromosome-positive cells. Autophagy 2016;12(6):936–48.

[101] Altman BJ, Jacobs SR, Mason EF, Michalek RD, Macintyre AN, Coloff JL, et al. Autophagy is essential to suppress cell stress and to allow BCR-Abl-mediated leukemogenesis. Oncogene 2010;30(16):1855–67.

[102] Yeo SK, Wen J, Chen S, Guan JL. Autophagy differentially regulates distinct breast cancer stem-like cells in murine models via EGFR/Stat3 and Tgfbeta/Smad signaling. Cancer Res 2016;76(11):3397–410.

[103] Sharif T, Martel E, Dai C, Kennedy BE, Murphy P, Clements DR, et al. Autophagic homeostasis is required for the pluripotency of cancer stem cells. Autophagy 2017;13(2):264–84.

[104] Robert T, Vanoli F, Chiolo I, Shubassi G, Berstein KA, Rothstein R, et al. HDACs link the DNA damage response, processing of double-strand breaks and autophagy. Nature 2011;471(7336):74–9.

[105] Avalos Y, Canales J, Bravo-Sagua R, Criollo A, Lavandero S, Quest AF. Tumor suppression and promotion by autophagy. Biomed Res Int 2014;2014, 603980.

124 Inflammation and oral cancer

[106] Aita VM, Liang XH, Murty VV, Pincus DL, Yu W, Cayanis E, et al. Cloning and genomic organization of beclin 1, a candidate tumor suppressor gene on chromosome 17q21. Genomics 1999;59(1):59–65.

[107] Liang XH, Jackson S, Seaman M, Brown K, Kempkes B, Hibshoosh H, et al. Induction of autophagy and inhibition of tumorigenesis by beclin 1. Nature 1999;402(6762):672–6.

[108] Huang Y, Chen L, Guo L, Hupp TR, Lin Y. Evaluating DAPK as a therapeutic target. Apoptosis 2014;19(2):371–86.

[109] Khan KH, Blanco-Codesido M, Molife LR. Cancer therapeutics: targeting the apoptotic pathway. Crit Rev Oncol Hematol 2014;90(3):200–19.

[110] Galluzzi L, Kepp O, Kroemer G. Mitochondria: master regulators of danger signalling. Nat Rev Mol Cell Biol 2012;13(12):780–8.

[111] Pattingre S, Tassa A, Qu X, Garuti R, Liang XH, Mizushima N, et al. Bcl-2 antiapoptotic proteins inhibit Beclin 1-dependent autophagy. Cell 2005;122(6):927–39.

[112] Miracco C, Cosci E, Oliveri G, Luzi P, Pacenti L, Monciatti I, et al. Protein and mRNA expression of autophagy gene Beclin 1 in human brain tumours. Int J Oncol 2007;30(2):429–36.

[113] Laddha SV, Ganesan S, Chan CS, White E. Mutational landscape of the essential autophagy gene BECN1 in human cancers. Mol Cancer Res 2014;12(4):485–90.

[114] Kroemer G, Levine B. Autophagic cell death: the story of a misnomer. Nat Rev Mol Cell Biol 2008;9(12):1004–10.

[115] Zhao Y, Yang J, Liao W, Liu X, Zhang H, Wang S, et al. Cytosolic FoxO1 is essential for the induction of autophagy and tumour suppressor activity. Nat Cell Biol 2010;12(7):665–75.

[116] Zhao Y, Wang L, Yang J, Zhang P, Ma K, Zhou J, et al. Anti-neoplastic activity of the cytosolic FoxO1 results from autophagic cell death. Autophagy 2010;6(7):988–90.

[117] Takamura A, Komatsu M, Hara T, Sakamoto A, Kishi C, Waguri S, et al. Autophagy-deficient mice develop multiple liver tumors. Genes Dev 2011;25(8):795–800.

[118] Nelson DA, Tan TT, Rabson AB, Anderson D, Degenhardt K, White E. Hypoxia and defective apoptosis drive genomic instability and tumorigenesis. Genes Dev 2004;18(17):2095–107.

[119] Papandreou I, Lim AL, Laderoute K, Denko NC. Hypoxia signals autophagy in tumor cells via AMPK activity, independent of HIF-1, BNIP3, and BNIP3L. Cell Death Differ 2008;15(10):1572–81.

[120] Adhauliya N, Kalappanavar AN, Ali IM, Annigeri RG. Autophagy: a boon or bane in oral cancer. Oral Oncol 2016;61:120–6.

[121] Azad MB, Chen Y, Gibson SB. Regulation of autophagy by reactive oxygen species (ROS): implications for cancer progression and treatment. Antioxid Redox Signal 2009;11(4):777–90.

[122] Li L, Chen Y, Gibson SB. Starvation-induced autophagy is regulated by mitochondrial reactive oxygen species leading to AMPK activation. Cell Signal 2013;25(1):50–65.

[123] Zoncu R, Efeyan A, Sabatini DM. mTOR: from growth signal integration to cancer, diabetes and ageing. Nat Rev Mol Cell Biol 2011;12(1):21–35.

[124] Frisch SM, Francis H. Disruption of epithelial cell-matrix interactions induces apoptosis. J Cell Biol 1994;124(4):619–26.

[125] Fung C, Lock R, Gao S, Salas E, Debnath J. Induction of autophagy during extracellular matrix detachment promotes cell survival. Mol Biol Cell 2008;19(3):797–806.

[126] Debnath J. p66(Shc) and Ras: controlling anoikis from the inside-out. Oncogene 2010;29(41):5556–8.

[127] Kenific CM, Debnath J. Cellular and metabolic functions for autophagy in cancer cells. Trends Cell Biol 2015;25(1):37–45.

[128] Qiang L, Zhao B, Ming M, Wang N, He TC, Hwang S, et al. Regulation of cell proliferation and migration by p62 through stabilization of Twist1. Proc Natl Acad Sci U S A 2014;111(25):9241–6.

[129] Egan DF, Chun MG, Vamos M, Zou H, Rong J, Miller CJ, et al. Small molecule inhibition of the autophagy kinase ULK1 and identification of ULK1 substrates. Mol Cell 2015;59(2):285–97.

[130] Petherick KJ, Conway OJ, Mpamhanga C, Osborne SA, Kamal A, Saxty B, et al. Pharmacological inhibition of ULK1 kinase blocks mammalian target of rapamycin (mTOR)-dependent autophagy. J Biol Chem 2015;290(48):28726.

[131] Dyczynski M, Yu Y, Ctrocka M, Parpal S, Braga T, Harley AB, et al. Targeting autophagy by small molecule inhibitors of vacuolar protein sorting 34 (Vps34) improves the sensitivity of breast cancer cells to Sunitinib. Cancer Lett 2018;435:32–43.

[132] Zahedi S, Fitzwalter BE, Morin A, Grob S, Desmarais M, Nellan A, et al. Effect of early-stage autophagy inhibition in BRAF(V600E) autophagy-dependent brain tumor cells. Cell Death Dis 2019;10(9):679.

[133] Kurdi A, Cleenewerck M, Vangestel C, Lyssens S, Declercq W, Timmermans JP, et al. ATG4B inhibitors with a benzotropolone core structure block autophagy and augment efficiency of chemotherapy in mice. Biochem Pharmacol 2017;138:150–62.

[134] Fu Y, Hong L, Xu J, Zhong G, Gu Q, Gu Q, et al. Discovery of a small molecule targeting autophagy via ATG4B inhibition and cell death of colorectal cancer cells in vitro and in vivo. Autophagy 2019;15(2):295–311.

[135] Mauthe M, Orhon I, Rocchi C, Zhou X, Luhr M, Hijjlkema KJ, et al. Chloroquine inhibits autophagic flux by decreasing autophagosome-lysosome fusion. Autophagy 2018;14(8):1435–55.

[136] Jensen M, Mehlhorn H. Seventy-five years of Resochin in the fight against malaria. Parasitol Res 2009;105(3):609–27.

[137] Finbloom DS, Silver K, Newsome DA, Gunkel R. Comparison of hydroxychloroquine and chloroquine use and the development of retinal toxicity. J Rheumatol 1985;12(4):692–4.

[138] Bedoya V. Effect of chloroquine on malignant lymphoreticular and pigmented cells in vitro. Cancer Res 1970;30(5):1262–75.

[139] Briceno E, Reyes S, Sotelo J. Therapy of glioblastoma multiforme improved by the antimutagenic chloroquine. Neurosurg Focus 2003;14(2), e3.

[140] Kroemer G, Galluzzi L. Lysosome-targeting agents in cancer therapy. Oncotarget 2017;8(68):112168–9.

[141] Rebecca VW, Nicastri MC, McLaughlin N, Fennelly C, McAfee Q, Ronghe A, et al. A unified approach to targeting the yysosome's degradative and growth signaling roles. Cancer Discov 2017;7(11):1266–83.

[142] McAfee Q, Zhang Z, Samanta A, Levi SM, Ma XH, Plao S, et al. Autophagy inhibitor Lys05 has single-agent antitumor activity and reproduces the phenotype of a genetic autophagy deficiency. Proc Natl Acad Sci U S A 2012;109(21):8253–8.

Chapter 7

Oral mucosal graft-versus-host disease and its possibility of antitumor effects

Kei Seno[a], Madoka Yasunaga[b], Nana Mori-Yamamoto[c], and Jun Ohno[d]

[a]Section of General Dentistry, Department of General Dentistry, Fukuoka Dental College, Fukuoka, Japan, [b]Section of Orthodontics, Department of Oral Growth and Development, Fukuoka Dental College, Fukuoka, Japan, [c]Section of Periodontology, Department of Odontology, Fukuoka Dental College, Fukuoka, Japan, [d]Oral Medicine Research Center, Fukuoka Dental College, Fukuoka, Japan

1. Introduction

Graft-versus-host disease (GVHD) is characterized by selective epithelial inflammation affecting the mucocutaneous organs, digestive tract, and the liver [1]. Among the mucocutaneous organs, oral mucosa is one of the tissues affected by GVHD [2]. Oral mucosal acute GVHD (aGVHD) is immunopathologically characterized by migration of effector cells into the surface epithelium, in which the events are mediated by intercellular adhesion molecule-1 (ICAM-1)/ lymphocyte adhesion function-associated antigen-1 (LFA-1), and others, via M1 macrophage infiltration. In this chapter, we will focus on the characteristics of GVHD and pathophysiological mechanisms of GVHD development, including oral mucosal aGVHD, as well as the possibility of GVHD-related antitumor effects (graft-versus-tumor, GVT).

2. What is GVHD?

GVHD is still a major complication and the leading cause of mortality after allogeneic hematopoietic stem cell transplantation (HSCT), which develops when transplanted donor T cells mount an alloimmune response against recipient (host) tissues, resulting in tissue damage [3]. Both acute and chronic complications ensue, often involving multiple organ systems. Formerly, acute GVHD (aGVHD) was defined as occurring prior to day 100, whereas chronic GVHD (cGVHD) occurred after that time [4–6]. As this definition is far from satisfactory—because aGVHD may develop after day 100—a recent classification

Inflammation and Oral Cancer. https://doi.org/10.1016/B978-0-323-88526-3.00007-5
Copyright © 2022 Elsevier Inc. All rights reserved.

includes late-onset aGVHD (after day 100) and an overlap syndrome with findings of both acute and cGVHD [7]. Bellingham first formulated three pathophysiological requirements for the development of GVHD: (i) the graft must contain immunologically competent cells; (ii) the recipient must express tissue antigens that are not present in the transplant donor; (iii) the recipient must be incapable of mounting an effective response to eliminate the transplanted cells [8]. Each element plays a crucial role in the development of both acute and chronic GVHD.

2.1 aGVHD

The classically recognized organs linked to aGVHD in humans are the skin, gastrointestinal tract, liver, and lymphatic organs (Fig. 1) [3]. The onset of aGVHD is usually marked by a maculopapular rash, often involving palms, soles, and ears. The characteristic maculopapular rash is pruritic and can spread throughout the body, sparing the scalp. These events may be associated with higher body temperature. In more severe cases, bullae can develop via epidermal-dermal separation.

Gastrointestinal tract involvement in aGVHD usually presents as diarrhea and nausea, sometimes abdominal pain, or even paralytic ileus [9]. Gastrointestinal bleeding, leading to poor prognosis, occurs as a result of mucosal ulceration [10]. Liver dysfunction may develop within days but more commonly within 2–4 weeks of transplantation. Generally, a gradual rise of both direct and indirect bilirubin is observed in association with a rise in alkaline phosphates and transaminases (AST and ALT). The liver may be enlarged, but patients rarely complain of pain. Histopathologically, the skin shows basal cell vacuolar degeneration and apoptosis, dyskeratosis, spongiosis, eosinophilic body formation of epidermal cells, and focal epidermal-dermal separation, occasionally progressing to complete epidermal loss, which is caused by epidermotropism of lymphocytes. Histological features in the gastrointestinal tract include patchy ulcerations, apoptotic bodies in the base of crypts, crypt

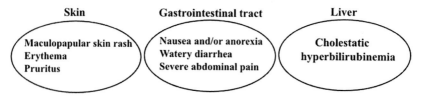

FIG. 1 Schematic overview of acute GVHD symptoms.

abscesses, and flattening, as well as complete loss of the surface epithelium [11]. Liver pathology of aGVHD includes eosinophilic necrosis of hepatocytes. Although other tissues may also be involved, generally only involvement of skin, intestinal tract, and liver is used for clinical grading of aGVHD. The overall clinical grading of aGVHD is classified as I (mild), II (moderate), III (severe), and IV (very severe). With grade I, only the skin is involved, with the symptom being a rash. With more extensive involvement of the skin, or involvement of liver or gastrointestinal tract, and impairment of the clinical performance status, either alone or in any combination, the overall grade advances from II to IV. Severe aGVHD carries a poor prognosis; with 25% long term survival for grade III and 5% for grade IV [12]. Furthermore, mortality is usually related to infectious complications as a result of the profound immunodeficiency associated with GVHD and also the immunosuppressive characteristic of its therapy. Patient death may occasionally be caused by hepatic failure or an abdominal catastrophe, such as bowel perforation or intestinal hemorrhage.

2.2 cGVHD

cGVHD continues to be a significant complication of allogeneic HCT and remains the leading cause of nonrelapse mortality. In the past 10 years, the incidence of aGVHD has remained constant, while that of cGVHD appears to have increased [13]. cGVHD is initially defined as a GVHD syndrome presenting more than 100 days after allogeneic HCT, either as an extension of aGVHD (progressive onset), after a disease-free interval (quiescent), or with no precedent (de novo) [5]. The clinical characteristics of cGVHD are similar to those of other immune mediated diseases such as lichen planus, lupus erythematosus, or systemic sclerosis [7]. In addition to target organs of aGVHD, symptoms of cGVHD are observed in other organs, such as nails, eyes, mouth, female genitalia, lung, kidney, heart, bone marrow, muscle fascia, and joints (Fig. 2). cGVHD incidence ranges from 30% to 50% of all allogeneic transplant patients, while aGVHD is observed in 50%–70% [14]. At present, there are no criteria allowing for the distinction between these two forms of GVHD, other than the clinical characteristics of each case. The acute form of GVHD is potentially fatal and typically affects the skin, gastrointestinal tract, and liver, whereas in cGVHD, the oral cavity is one of the most commonly affected regions, and may be the sole body location affected by the disease [15]. Recently, proposed diagnostic approaches differentiating between acute and cGVHD have been postulated [7].

3. Oral GVHD

Oral findings are observed in both acute and chronic GVHD.

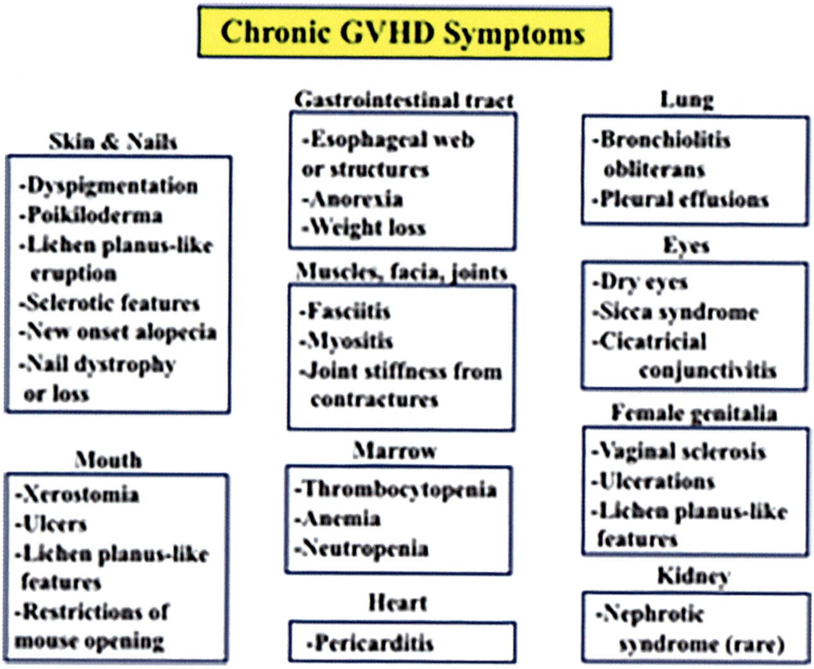

130 Inflammation and oral cancer

Chronic GVHD Symptoms

Skin & Nails
- Dyspigmentation
- Poikiloderma
- Lichen planus-like eruption
- Sclerotic features
- New onset alopecia
- Nail dystrophy or loss

Gastrointestinal tract
- Esophageal web or structures
- Anorexia
- Weight loss

Muscles, facia, joints
- Fasciitis
- Myositis
- Joint stiffness from contractures

Lung
- Bronchiolitis obliterans
- Pleural effusions

Eyes
- Dry eyes
- Sicca syndrome
- Cicatricial conjunctivitis

Mouth
- Xerostomia
- Ulcers
- Lichen planus-like features
- Restrictions of mouse opening

Marrow
- Thrombocytopenia
- Anemia
- Neutropenia

Heart
- Pericarditis

Female genitalia
- Vaginal sclerosis
- Ulcerations
- Lichen planus-like features

Kidney
- Nephrotic syndrome (rare)

FIG. 2 Schematic overview of chronic GVHD symptoms.

3.1 Oral aGVHD

Although prevalence rates of oral aGVHD remain unclear, oral features developing systemic aGVHD have been reported infrequently [2, 16–23]. Woo et al. [2] speculated that approximately 60%–70% of patients with grade III to grade IV aGVHD may have oral changes consistent with GVHD. The clinical diagnosis of oral aGVHD was made at the time when oral lesions occurred concomitantly with systemic aGVHD [17, 20]. Oral aGVHD presents generalized mucosal erythema and painful desquamative and ulcerative lesions, which involve almost the entire oral cavity. Among those symptoms, general lesions are characterized as extensive, irregular nonspecific erythema, and ulcerations of the keratinized and nonkeratinized mucosa. Clinically, oral aGVHD is defined at the onset of oral mucosal erythema and ulcerations in the context of engraftment and aGVHD involvement of typical target organs. Biopsy analysis of oral lesions is not performed in all cases, particularly when the diagnosis of systemic aGVHD is already established. In patients who receive conditioning regimen to HSCT, oral ulcerations may develop before the onset of oral aGVHD lesions.

Management of oral mucosal aGVHD is directed toward pain control and local measures, in addition to management of systemic GVHD. In particular, topical and local management of symptomtic oral aGVHD, using steroid rinses and gels, is necessary because severe and intractable ulcerations can interfere with oral function and quality of life. To date, optimal management of oral mucosal aGVHD has not been established. Berkowtiz et al. [22] reported that the oral lesions of aGVHD typically resolve rapidly after initiation of systemic corticosteroid therapy. Topical steroid application appeared to help reduce the severity and duration of oral lesions associated with aGVHD [21]. However, effectiveness of both systemic and topical application of steroids remained unclear. Due to the absence of available guidelines for GVHD management, the principles of oral aGVHD management were generally followed [15]. For instance, most patients were treated with topical high-potency corticosteroids (dexamethasone or compounded solutions) and tacrolimus (ointment and compound solution), which seemed to provide clinical benefit. In cases of severe pain that was uncontrolled by systemic opioids, topical morphine solution appeared to be effective in managing symptoms [24, 25].

3.2 Oral cGVHD

In patients with cGVHD, the oral cavity is affected in 70% of those who have undergone HSCT and in 35% of those who have received a bone marrow transplant. The prevalence of oral cGVHD ranges from 45% to 83%, and the oral cavity, together with the skin, is one of the main target organs of cGVHD, though manifestations are also commonly seen in the lungs, liver, genitals, and gastrointestinal tract [26]. Oral cGVHD is usually classified as a singular clinical entity, but, in reality, there are unique clinical features associated with mucosal, salivary gland, and sclerodermatous involvement, each of which contribute to significant morbidity and late complications. In oral mucosal cGVHD, the lesion is characterized by lichenoid inflammation that may appear in all intraoral sites, but which particularly affects the tongue and buccal mucosa [15, 16, 27]. Clinical findings of oral mucosal cGVHD include white hyperkeratotic reticulations and plaques, erythematous changes and ulcerations, ranging from limited lesion with only mild reticulation to more extensive disease with painful ulcerations. Erythema and ulcerations are overlap symptoms between oral acute and chronic GVHD. Symptoms of cGVHD rarely extend posteriorly to the oropharynx because the soft palate is infrequently affected. The lips, one of the frequent sites of cGVHD involvement, demonstrate the same changes observed in oral mucosal cGVHD, providing a source of significant discomfort. In patients affected by prominent reticular changes in the buccal mucosa, a sensation of mouth tightness and a reduced ability to open the mouth may be reported. These symptoms must be distinguished from oral tightness because of primary sclerotic cutaneous cGVHD or secondary to mucosal scarring [15].

Although oral mucosal cGVHD presents with prominent features that are easily recognized clinically, involvement of the salivary glands tends to be less obvious. In addition, some patients treated with conditioning regimen of HSCT

132 Inflammation and oral cancer

(usually associated with total body irradiation or prior irradiation to the head and neck) demonstrate xerostomia (subjective complaint of oral dryness). This condition may persist through the period when salivary gland cGVHD develops, making onset and diagnosis less evident. Salivary gland cGVHD results in both qualitative and quantitative changes in salivary production, composition, and output. Extraoral salivary gland swelling, with or without discomfort, may be secondary to obstructive changes but is exceedingly rare. Results of intraoral examination may demonstrate dry-appearing oral mucosa with lack of floor-of-mouth pooling. Patients with salivary gland cGVHD are at risk of developing secondary infectious complications because anticariogenic and antifungal activities are diminished. Furthermore, patients are at significant risk of recurrent oral candidiasis, especially if there is ongoing topical corticosteroid therapy for management of oral mucosal cGVHD, suppressing mucosal immunity. The development of recurrent superficial mucoceles is an additional feature of oral cGVHD. Mucoceles generally consist of painless mucus-filled blisters that develop primarily on the palate but can also be seen on the labial and buccal mucosa and tongue, or wherever there are minor salivary glands. These superficial mucoceles can be found in the context of both mucosal and salivary gland cGVHD, suggesting that the underlying inflammation may be secondary to generalized mucosal involvement or due to direct salivary gland tissue targeting.

cGVHD-induced sclerotic changes are relatively rare in the oral cavity, but when present, may be associated with significant morbidity. Signs and symptoms of sclerotic changes include limited mouth opening, pain, secondary ulceration, and impaired oral hygiene. Involvement results in perioral and facial skin sclerosis, typically as an extension of more generalized sclerotic changes. Primary mucosal sclerosis, developed as a sequela of long-standing severe ulcerative mucosal cGVHD, causes band-like fibrosis in the posterior buccal mucosa. In other associated conditions of oral cGVHD, herpes simplex virus infection can present as solitary or multiple ulcerative lesions affecting any intraoral sites.

As oral cGVHD and its treatment are associated with different complications, such as secondary infections, osteoporosis, hypertension, hyperglycemia, renal failure, and hyperlipidemia, no specific drug therapy has yet been approved. The primary treatment goals are to diminish patient pain, maintain the patient's ability to eat, increase the patient's quality of life, and prevent destruction of oral tissues and dentition. Pharmacotherapy for oral cGVHD may be systemic, topical, or injectable [15, 27]. Systemic immunosuppressive therapy is indicated in extensive cGVHD involving multiple organs or body sites. The primary limitation of systemic immunosuppression is the increased risk for opportunistic infections, which is a leading cause of mortality in HSCT patients. Cyclosporine and systemic corticosteroids, either alone or in combination, are commonly used as systemic immunosuppressive drugs. Cyclosporine, a calcineurin inhibitor, suppresses T-cell proliferation and prevents transcription of genes for IL-2, IL-2 receptor, and IFN-γ. Patients taking cyclosporine are at increased risk of oral mucosal infections, such as oral candidiasis and herpetic infections. Systemic corticosteroids also predispose patients to fungal and viral infections. Rituximab, an anti-CD20 monoclonal antibody, has demonstrated promise in the treatment of cGVHD [28, 29]. The drug was effective in over 50% of patients with refractory cGVHD and may have a beneficial impact upon survival. Topical and local therapy for oral cGVHD has several advantages, including fewer systemic side effects and drug interactions, the ability to intensify therapy to one specific area while preventing systemic host immune-suppression, and maintenance of graft-versus-host antitumor effects. Topical corticosteroids, commonly used for many oral mucosal inflammatory conditions, are the most popular local therapy for oral cGVHD. Agents tested include topical budesonide rinse [30] and topical dexamethasone rinse [31]. As decreased salivary flow is a major complain of patients with salivary gland cGVHD, treatment for increased salivary flow is necessary to increase patient comfort [32]. Patients with reduced salivary flow have an increased risk of dental caries and oral fungal infections. Agents that increase patient comfort include cholinergic agonists pilocarpine which increase salivary flow rates.

4. Immunopathophysiology of oral GVHD

4.1 Contribution of animal models for pathophysiology of GVHD

The use of animal models to study human diseases is considered essential for understanding underlying pathophysiological and molecular mechanisms [33]. Barnes and Loutit demonstrated that, when irradiated mice were infused with allogenic marrow and spleen cells, they recovered from radiation injury and aplasia but developed diarrhea, weight loss, skin changes, and liver abnormalities, subsequently dying due to "secondary disease" [34]. This phenomenon was recognized as GVHD. GVHD is an excessive inflammatory reaction caused by the donor-source immune competent cells attacking the host cells and organs [9]. It is a complex pathophysiological process because this disease is represented by an immunoreaction mainly triggered by T cells: tissue damage caused by clone proliferation of the donor-source T cells after contacting and identifying the antigens of the recipient, or the T cells identifying the major histocompatibility complex (MHC) of the recipient's target cells directly, or secreting cytokines directly.

Currently, most of our understanding of GVHD pathogenesis is obtained from animal models. Furthermore, characterization of the immune competent cells in GVHD patients is only conducted *in vitro*. As the results of *in vitro* studies are far from those obtained *in vivo*, the stable and reliable approaches of humanized animal models are crucial for advancing GVHD studies. Frequently, MHC-mismatched donors are used in GVHD models. The aggressive nature of GVHD follows the transfer of small numbers of parental strain ($H2^x$) T cells into a conditioning regimen (irradiation or chemotherapy)-treated F1(H2^{xy}) mice. This may give rise to a severe condition because the host expresses combined H2 class I and II differences, mimicking the fact that GVHD involves CD8$^+$ or CD4$^+$ T cells, responding to MHC class I and II differences, respectively. HSCT in humans is primarily restricted to donor/host combinations between human leukocyte antigen (HLA)-compatible siblings or matched unrelated donors; therefore, murine models for GVHD directed to minor H antigens (miHA) are of obvious clinical relevance. GVHD in HLA-compatible combinations can be very severe. P-F1 GVHD models are performed with nonirradiated hosts. In these models, large doses of either donor CD4$^+$ or CD8$^+$ T cells result in lethal GVHD within 2–3 weeks. This GVHD is characterized by prominent lymphocytic infiltrations in various target organs and splenomegaly, which is pronounced by about 10 days postinjection [35–38].

4.2 A three step model of aGVHD pathophysiology

The pathophysiology of aGVHD can be attributed to a three step process where the innate and adaptive immune systems interact [39]. These steps are: (1) tissue damage due to the radiation/chemotherapy pretreatment conditioning regimen (this process, in turn, activates the host antigen-presenting cells (APCs)); (2) an afferent phase to activate donor T cells; and (3) an efferent phase in which cellular and inflammatory factors work together to damage the target organs (Fig. 3).

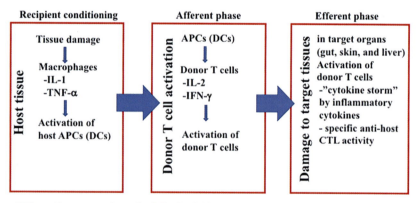

FIG. 3 The pathophysiology of acute GVHD. This process is comprised of three steps: (1) tissue damage due to pretreatment conditioning regimen, (2) afferent phase: activation of donor T cells, and (3) efferent phase: destruction of target tissues. APCs, antigen-presenting cells; DCs, dendritic cells; IL-1, interleukin-1; TNF-α, tumor necrosis factor-α; IL-2, interleukin-2; IFN-γ, interferon-γ; CTL, cytotoxic T lymphocyte.

The first step of aGVHD occurs before infusion of donor cells. Prior to HSCT, the conditioning regimen (irradiation and/or chemotherapy) leads to damage and activation of host tissues throughout the body. Damaged host tissue releases proinflammatory cytokines, such as tumor necrosis factor (TNF)-α and interleukin (IL)-1 [40], that activate host APCs, ultimately activating donor T cells present in the stem cell inoculum [41, 42]. These cytokines may provide enhancement of donor T-cell recognition of host alloantigens by increasing expression of MHC antigens and other molecules on host APCs. Proinflammatory cytokines may also stimulate chemokine release, recruiting donor T cells into host target organs. Chemotherapy as the conditioning regimen can activate host APCs and help to recruit donor T cells infused in the stem cell inoculum. Total body irradiation also activates host tissues to secrete inflammatory cytokines, such as TNF-α and IL-1 [40], and elicits endothelial apoptosis leading to epithelial cell damage in the gastrointestinal tract [43].

Donor T-cell activation occurs during the second step of aGVHD. In murine studies, host APCs alone are both necessary and sufficient to stimulate donor T cells [41, 44]. Activated donor T cells can recognize alloantigen either on host or donor APCs, known as direct or indirect antigen presentation, respectively [45, 46]. In the direct presentation, donor T cells recognize either the peptide bound to allogeneic MHC molecules or the foreign MHC molecules themselves [45]. During the indirect presentation, T cells respond to the peptides generated by degradation of the allogeneic MHC molecules that are presented on self MHC [47]. Costimulatory signals are required for further activation of APCs, which in turn enhance T-cell stimulation, characterized by continuous proliferation of T cells and differentiation into types Th1 and Th17. Differentiated T cells are involved in the activation of CD4 cytotoxic T lymphocyte (CTL), CD8 CTL, and natural killer cells that mediate tissue damage. The majority of Th1 cells secrete IFN-γ, which goes on to activate mononuclear phagocytes.

In the third phase (the efferent phase), both innate and adaptive immune responses work synergistically to make the T cell-induced inflammation worse. This phase includes (1) inflammatory cytokines, (2) specific antihost CTL activity using Fas and perforin pathways, (3) large granular lymphocytes or NK cells, and (4) nitric oxide (NO). Cellular mediators, such as CTLs and NK cells, utilize the Fas/Fas ligand pathway and perforin/granzyme pathway to lyse the target cells [48, 49]. Perforin is stored together with granzymes and other proteins in cytotoxic granules in CTLs and NK cells [50]. In the effector phase of aGVHD, inflammatory cytokines, in association with CTLs, enhance the amplification of local tissue injury and the development of target organ dysfunction in the transplant recipient. Cytokines TNF-α and IL-1 are produced by an abundance of cell types involved in innate and adoptive immune responses, and they have synergistic and redundant roles during several phases of aGVHD. High levels of TNF-α mRNA were noted in GVHD tissues and target organ damage could be inhibited by infusion of anti-TNF-α mAbs, reducing mortality from 100% to 50% by the administration of the soluble form of the TNF-α receptor

136 Inflammation and oral cancer

[40, 51, 52]. Interestingly, TNF-α plays a central role in intestinal and cutaneous as well as lymphoid GVHD [52–55], while it is less important in hepatic GVHD [44, 56]. IL-1 is the second major proinflammatory cytokine that contributes to aGVHD damage. Secretion of IL-1 appears to occur predominantly during the effector phase of aGVHD in the spleen and skin, known to be two major aGVHD target organs [57]. Similarly, an increase in IL-1 mRNA from mononuclear cells has been shown during clinical aGVHD [58]. Significant amounts of NO can be produced by macrophages as a result of activation during GVHD. Several studies have demonstrated that NO contributes to the deleterious effects of GVHD on target tissues and specifically to GVHD-induced immunosuppression [59, 60]. NO also inhibits proliferation of epithelial stem cells in the gut and skin, resulting in inhibition of repair mechanisms of target tissues [61]. Effector functions of mononuclear phagocytes and neutrophils are facilitated through a secondary signal (provided by microbial products like lipopolysaccharide (LPS)), that leak through the damaged intestinal mucosa and skin and are released by HSCT conditioning regimens during the first step.

4.3 Pathophysiology of mucocutaneous aGVHD

The histopathology of aGVHD is characterized by selective inflammation affecting the mucocutaneous organs, digestive tract, and liver [1]. As described above, the oral mucosa is one of the targets for mucocutaneous GVHD [2]. The cutaneous effects of GVHD were originally stated as a "secondary disease" in a murine model of HSCT in 1955, which distinguished them from the primary toxicity of radiation sickness [62]. Since then, many studies have been conducted to define the pathology and increase understanding of cutaneous and/or mucocutaneous aGVHD.

The first histological changes observed in mucocutaneous aGVHD involve adhesion and transvascular diapedesis of lymphocytes across postcapillary venues situated within the uppermost dermis (the skin) or lamina propria (the oral mucosa), in association with the first signs and symptoms of mucocutaneous involvement [63]. Activation of microvascular endothelial cells, which is responsible for initial recruitment of effector cells, appears to be confirmed by mast-cell degradation around the affected venules. As this endothelial phase of aGVHD is nonspecific, it is of less value for diagnosis. Lymphocytes migrate from the perivascular interstitium into the overlying epidermal layer (epidermotropism). Some lymphocytes may align along the dermal epidermal junction, while others are present in all levels of the epidermis. Therefore, finding lymphocytes in the epidermis is a valuable diagnostic tool for early mucocutaneous aGVHD.

Mucocutaneous aGVHD, unlike other dermatitides showing lymphoid epidermotropism, generally does not reveal significant intracellular edema (spongiosis) within the epidermal layer. The most specific histological finding in mucocutaneous aGVHD is "satellitosis," indicating target cell damage within the epidermis. Satellitosis occurs when multiple lymphocytes intimately surround one or several keratinocytes (KCs) that reveal eosinophilic degeneration and cell death. KCs affected by satellitosis usually contain condense, hyper chromatic, and sometimes fragmented nuclei. Recent evidence indicates that such dying target KCs are actually undergoing apoptosis [55]. During this process, lymphocytes, particularly CD8$^+$ cells, migrate from the perivascular interstitium into the overlying epidermal layer and induce degenerative changes in KCs. In oral squamous mucosal epithelium, aGVHD is characterized by a band-like infiltrate of lymphocytes directly beneath the surface epithelial layer. Like cutaneous aGVHD, lymphocyte infiltrate into the epithelium may be associated with apoptosis and satellitosis.

These histopathological findings remind us once again that aGVHD is a cytotoxic attack of donor lymphocytes on host tissues, most dramatically on epithelial cells. Wagner and Murphy have conceptually divided this process in mucocutaneous aGVHD into three distinct phases: allostimulation, homing, and targeting [63]. The first step of mucocutaneous aGVHD pathophysiology includes allostimulation and the initiation of cytokine cascades. The basic mucocutaneous lesion triggers the initial influx of specifically allostimulated T cells. The donor T cells mediating mucocutaneous aGVHD are directed toward HLA or minor histocompatibility antigen (miHA) disparities expressed by host tissues [9, 64]. Experimental evidence implicates alloreactive responses occurring in secondary lymphoid tissues, such as Peyer's patches in the gut [65]. Specifically allostimulated T cells promote the development of mucocutaneous lesions by infiltrating and damaging host tissues. Accordingly, identification of the specific alloreactive T cells related to the induction of aGVHD is important in defining the histopathology of the earliest mucocutaneous lesions. In murine GVHD models, specific subsets of allostimulated cells have been identified by the T-cell receptor (TCR) Vβ repertoire of CD4$^+$ T cells. Expansion of specific Vβ families has been documented in a murine model of miHA-induced aGVHD [66]. Furthermore, using antibodies specific for these Vβ populations, skin and squamous mucosal infiltration by these allostimulated cells can be documented. The homing patterns of these effector T cell subpopulations are likely to provide insight into the identification of cellular targets within tissues.

Cytokines lead the development and formation of mucocutaneous aGVHD. In addition to the effects of HSCT conditioning regimen [67], cytokine release is responsible for aggravating this condition as a consequence of transplantation and allostimulation. The release of these inflammatory cytokines in mucocutaneous aGVHD has even promoted the concept of GVHD as a "cytokine storm" [68]. The outcome of systemic and local cytokine release leads to cellular activation, induction of cell adhesion pathways that mediate organ-specific homing, and direct cytotoxicity. Some important cytokines released during the progression of mucocutaneous aGVHD include IL-1 [57, 68–70], IL-2 [57, 68, 69], IFN-γ [71], TNF-α [51, 52, 57], and transforming growth factor-beta (TGFβ) [72, 73]. Activation of specific cytokine cascades during aGVHD occurs in three phases [67]. In the first phase, inflammatory cytokines TNF-α and IL-1 are secreted because the conditioning regimen results in activation and damage to host tissue. In the second phase, donor T-cell activation elicits proliferation of Th1 T cells and secretion of IL-2 and IFN-γ, which in turn induce further T-cell expansion, cytotoxic T lymphocyte (CTL), NK cell responses, and prime mononuclear phagocytes to produce additional IL-1 and TNF-α. In the last phase, effector functions of aGVHD are triggered by LPS, which leaks through intestinal mucosa damaged initially by the conditioning regimen. Ultimately, LPS may stimulate gut-associated mononuclear cells as well as skin cells to release additional proinflammatory cytokines.

Although the great mystery of aGVHD is still why and how allostimulated effector T cells find their way to epithelial tissues in target organs, including mucocutaneous tissues, phase II events are explained as homing of effector cells into the target epithelium. The microanatomy of the superficial dermis (lamina propria, in the oral mucosa) is critical to understanding the cellular and molecular interactions related to target-specific leukocyte homing. In particular, the microvascular plexus, called the papillary dermis, is important for epidermal or epithelial homing of lymphocytes because this tissue consists of capillary loops extending upward within conical dermal papillae that invaginate into the epidermal layer. The perivascular microenvironment of post capillary venules contains mast-cell, monocyte/macrophages, and dermal dendritic cells. In the earliest phases of mucocutaneous inflammation, prominent alterations occur in the postcapillary venules and surrounding cells. These changes consist of degranulation of perivenular mast cells and endothelial activation. Release of histamine from mast cells leads to endothelial cell activation, characterized by prominence of endothelial cytoplasmic filaments, endothelial cell bulging into the venular lumen, and gap formation between adjacent endothelial cells.

Furthermore, this activation orchestrates molecular signals related to endothelial adhesion molecules that serve to promote leukocyte binding to the lumenal endothelial surface. P-selectin triggers circulating leukocytes to slowly "roll" along the endothelial membrane. Some molecules may contribute either to favor adhesion for specific subpopulations of mucocutaneous homing effector cells (e.g., E-selectin) or to strengthen leukocyte-endothelial binding and promoting diapedesis, such as vascular cell adhesion molecule-1 (VCAM-1), intercellular adhesion molecule-1 (ICAM-1), and platelet endothelial cell adhesion molecule-1 (PECAM-1 or CD31). Complementary ligands on effector cells may be constitutively expressed by certain subpopulations, such as lymphocyte function-associated antigen-1 (LFA-1), known as the ligand for ICAM-1 [74]. Secretory substances, mainly histamine, originated by mast-cell degranulation, may enhance these binding cascades either to induce or alter distribution of endothelial adhesion molecules, such as P-selectin, E-selectin, ICAM-1, and VCAM-1. Once effector T cells have accumulated in sufficient numbers in the papillary dermis, they finally navigate into the epidermal or epithelial layers, causing the cellular injury that typifies mucocutaneous aGVHD.

The third phase of mucocutaneous aGVHD pathophysiology, called targeting phase, consists of epidermotropism of effector cells and epithelial target cell injury. Epidermotropism is a process by which effector cells, after accumulating in the upper dermis, navigate into the epidermal (epithelial) layer. In the initial step of the epidermotropism, the adhesive events between effector cells and epidermal (epithelial) cells appear to occur in the basal cell layer and the basement membrane (BM) zone, which is defined as the boundary between epidermis (epithelium) and superficial dermis (lamina propria). Actually, many T cells that infiltrate the epidermal (epithelial) layer become located in close proximity to the BM zone. Those T cells, but not B cells, express the $\alpha3\beta1$ membrane integrin, which is a receptor for the BM ligand laminin 5 [75]. Laminin 5 is synthesized by basal keratinocytes (KCs) and is expressed primarily in the lamina lucida of the BM. Binding interactions between effector cells and basal KCs are therefore likely to influence early epidermotropic migration and adhesion in aGVHD.

After attachment of effector cells in the basal KCs, these cells migrate to the suprabasal epidermal (or epithelial) layers. Dislocation of effector cell to the suprabasal layer is likely to be mediated by interactions between leukocytes and KCs, including well-established LFA-1/ICAM-1-mediated adhesion. Immunohistochemical approaches in human and experimental aGVHD revealed that a predominance of infiltrating CD8[+] T cells and increased epidermal (epithelial) expression of ICAM-1 and MHC class II molecules are observed in the injured epidermis of the skin. As KCs do not constitutively express ICAM-1, unlike endothelial cells, expression by KCs must be induced by systemic cytokines as well as those secreted by early migrant T cells. Epidermal KCs are a major target of IFN-γ and TNF-α. IFN-γ and, to a lesser degree, TNF-α, stimulate the expression of ICAM-1 and MHC class II molecules by KCs *in vitro* and *in vivo* [76–78]. Therefore, IFN-γ-induced epidermal expression of ICAM-1 is prominently displayed during the epidermotropic phases of mucocutaneous aGVHD.

In the setting of experimental rat aGVHD, ICAM-1 expression in epithelial KCs is triggered at the onset of epithelial destruction by aGVHD-related responses in the oral mucosa [37]. Tissue expression of IFN-γ mRNA and expression of IFN-γ receptor by KCs were detected in early lesion of oral mucosal aGVHD, indicating that locally produced IFN-γ induced ICAM-1 expression by KCs. Using the Stamper-Woodruff binding assay, CD8[+] cells were observed bound to KCs in frozen sections of epithelial lesions, whereas no lymphocyte attachment was observed in normal KCs. Adherence could be inhibited by pretreating CD8[+] cells with LFA-1 antibody and/or by pretreating sections with ICAM-1 antibody, indicating that binding of effector cells in KCs is mediated in part by the LFA-1/LFA-1 pathway. These findings provide additional support for the role of ICAM-1, LFA-1, and IFN-γ in the development of oral mucosal aGVHD, by inducing migration of effector cells into the oral epithelium where they adhere to KCs. Furthermore, VCAM-1 expression was observed in the basal KCs close to the upper dermis, in which effector cells accumulated.

Effector T cells initiate and perpetuate epidermal (epithelial) injury in aGVHD after successful homing to skin or oral mucosa. These cells include NK cells, lymphokine-activated killer cells, and CTL. NK cells have been found in close association with dead or dying epithelial cells in the skin, liver, and colon, suggesting that these cells may act as possible effector cells, infiltrating organs of mice with aGVHD [79]. $CD8^+$ CTL are also known to play an important role in mucocutaneous aGVHD. $CD8^+$ cytotoxic T cells are at the onset of the acute phase of syngeneic GVHD in rats.

Recently, Seno, Yasunaga and coworkers [38] have postulated that macrophages contribute to the development of oral mucosal aGVHD, using a P-F1 nonirradiated GVHD model in rats. Macrophages are tissue-resident professional phagocytes and antigen-presenting cells, which differ from circulating peripheral blood monocytes. They are divided into two major populations, M1 and M2, based on *in vitro* data [80, 81]. The M1 subtype comprises classically activated macrophages stimulated by IFN-γ and lipopolysaccharide (LPS), whereas the M2 subtype comprises alternatively activated macrophages following stimulation with IL-4 and IL-13 [82]. At the onset of oral mucosal aGVHD, M1 macrophages accumulate in the BM via the laminin/CD29 β1 integrin pathway (Fig. 4). BM degradation was a result of the release of macrophage-secreted matrix metalloproteinase-2. The adhesion of M1 macrophages to the oral epithelial KCs could be inhibited by pretreating macrophages with a CCR2 (CC chemokine receptor 2) antibody and/or pretreating injured tissue sections with a monocyte chemoattractant protein-1 (MCP-1) antibody. These findings suggest that migration and adhesion of M1 macrophages, which is mediated in part by both lamina/CD29 β1 integrin and MCP-1/CCR2 pathways, is associated with the development of oral mucosal aGVHD.

Activated macrophages accumulate and persist in oral mucosa aGVHD

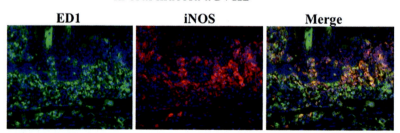

FIG. 4 Immunofluorescence images of ED1 and inducible nitric oxide synthase (iNOS) in oral mucosal aGVHD. ED1 (*green*) and iNOS (*red*) are pan and M1 macrophage markers, respectively. The nuclei were stained with Hoechst 33324 (*blue*). (Original magnification, x200. Adapted from Seno K, Yasunaga M, Kajiya H. et al., Dynamics of M1 macrophages in oral mucosal lesions during the development of acute graft-versus-host disease in rats. Clin Exp Immunol 2017;190(3):315–327).

During the progress of mucocutaneous aGVHD, cell injury in KCs is characterized by satellitosis, where signs of eosinophilic degeneration and cell death, and vacuolization of basal KCs, result in dermal epidermal detachment. These KC changes then induce apoptosis [55]. Molecular mechanisms responsible for the induction of apoptosis by infiltrating cytotoxic effector cells in mucocutaneous tissues [83, 84] involve one or more of at least three pathways, such as Fas/Fas ligand, perforin/granzyme [49, 85], and TNF/TNF receptor-tumor necrosis factor-related apoptosis-inducing ligand (TRAIL) [86, 87]. The Fas/Fas ligand pathway plays an important role in target cell cytolysis by CTL, regulation of inflammatory responses, peripheral deletion of autoimmune cells, costimulation of T cells, and activation-induced cell death, leading to regulatory mechanisms for mucocutaneous aGVHD. It has been suggested that donor T cells may mediate GVHD activity primarily through the Fas ligand effector pathway. Moreover, studies employing donor murine T cells deficient in TRAIL demonstrated diminished graft-vs-leukemia (GVL) but not GVHD effects, suggesting that TRAIL is also a crucial mediator of GVL [88].

KCs must undergo their molecular changes prior to direct injury or even before migration of effector cells into the epidermis (epithelium). There are many molecules that are expressed in KCs during the development of human and experimental mucocutaneous aGVHD, including ICAM-1 [37, 63, 89, 90], VCAM-1 [91], MHC class II [37, 89, 90, 92–94], IFN-γ receptor [37], MCP-1 [38], and mannose (Man)-specific *Lens culinaris* lectin (LCA; Fig. 5) [35]. In mucocutaneous aGVHD, inducible MHC class II expression is one of the features in KCs. Actually, immunohistochemical assays for MHC class II and ICAM-1 could be helpful in diagnosing early aGVHD. A study has provided the intriguing observation that ICAM-1 expression by KCs precedes MHC

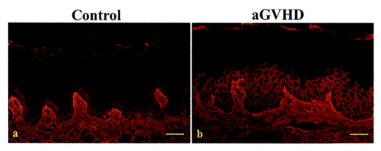

FIG. 5 Expression in oral mucosal aGVHD of *Lens culinaris* lectin (LCA), known as Man-binding lectin. (A) LCA binding on the cell surface of basal to paranasal cells in control mucosa. (B) In aGVHD, staining extends to the spinous layers of the surface epithelium. Bars = 100 μm. (Adapted from Hanada H, Ohno J, Seno K. et al. Dynamic changes in cell-surface expression of mannose in the oral epithelium during the development of graft-versus-host disease of the oral mucosa in rats. BMC Oral Health 2014;14(5):5).

142 Inflammation and oral cancer

class II expression in a rat aGVHD model [37]. During the early oral mucosal aGVHD, the basal KCs uniformly expressed ICAM-1 but not MHC class II molecules. This finding is consistent with parallel observations that ICAM-1 expression was strongly induced in cultured KCs after treatment with IFN-γ for 24 h, whereas MHC class II expression could not be detected, suggesting that increased expression of ICAM-I may be correlated with the onset of MHC class II expression in KCs.

Like some skin diseases, where the INFγ receptor can be expressed in KCs, this receptor is expressed in aGVHD-related oral epithelial KCs, providing indirect evidence that the local release of IFN-γ at the onset of aGVHD impacted KCs. A recent study has examined the migration and adhesion of M1 macrophages from oral mucosal aGVHD to epithelial KCs via the interaction of MCP-1/CCR2, suggesting that M1 macrophages may play an important role in the development of mucocutaneous aGVHD [38]. Although data pertaining to aGVHD-related molecules in KCs have accumulated, mechanisms for induction of aGVHD-related cell injury remain unclear. Further exploration of these issues is required in order to more precisely clarify the target KCs and molecules required for apoptotic injury in mucocutaneous aGVHD.

5. Graft-versus-tumor as antitumor effect of GVHD in oral squamous cell carcinoma

Oral squamous cell carcinoma (OSCC) is the most common malignant tumor in the maxillofacial region, making it the eleventh most common cancer worldwide [95, 96]. Advances in therapy have improved quality of life, but a poor survival rate of 50% or less at 5 years has remained unchanged over the past decades, regardless of the multidisciplinary treatment used [95, 96]. Mortality caused by OSCC remains high because of the development of distal metastases and the emergence of local and systemic recurrences resistant to chemo- and radiotherapies.

Studies in animal models and subsequent clinical experience in humans have shown graft-versus-tumor (GVT) effects, where donor T cells from the graft can eliminate residual malignant cells. In 1956, Barnes and colleagues observed that an allogeneic reaction of the bone marrow graft against the recipient's leukemia might contribute to a cure for leukemia [97]. This immunological event is called the "graft-vs-leukemia" (GVL) effect [98], and provides evidence for a role of the immune system in controlling human cancers. However, as the preferred reactions for an antileukemia effect are alloimmune responses, these responses are also responsible for the development of GVHD. The immunobiological principles of development of GVHD and GVT responses are highly similar. To succeed in establishing the GVT effect for antitumor therapy, separation between GVT and GVHD is required in clinical approaches.

The balance between GVT and GVHD depends on multiple factors. If donor T cells only recognize polymorphic antigens selectively expressed on cancer cells of the recipient, GVT may occur in the absence of severe GVHD. However, induction of immune responses against antigens only expressed on cancer cells may be difficult to achieve. A positive approach in inducing anti-cancer effects by alloimmune responses must be to reduce GVHD while preserving GVT reactivity. After T-cell-replete transplantation, activated recipient APCs and homeostatic proliferation of donor T cells will occur, resulting in high risk of GVT reactivity accompanied by GVHD. In contrast, abrogation of GVHD development is caused by complete removal of donor T cells but, as expected, GVT reactivity is also abrogated, resulting in an increased risk of relapse, high risk of infections, and lack of pathogen control. Recent studies have postulated that early sequential donor lymphocyte infusion (DLI) is likely necessary for successful GVT effect [99, 100]. Scheduled postponed unmanipulated DLI may restore GVT reactivity, with more limited risk of GVHD due to tissue repair, less inflammation, and replacement of recipient APC by donor APC.

Generally, GVT is a well-described phenomenon in patients with hematologic malignancies and serves as a major contributing factor in eliminating cancer cells. It has also been observed in various advanced solid malignancies including renal cell cancer [101–104], ovarian cancer [105, 106], breast cancer [107, 108], liver carcinoma [108, 109], and melanoma [110, 111]. In contrast to GVL activity, mechanisms of GVTs' effect in solid tumors remain unclear. Most of GVTs' effects in solid tumors have been reported in patients with acute or cGVHD, indicating that these effects could not be separated from the development of GVHD. This finding provides indirect evidence that donor effector T cells can be directed against both antigens restricted to normal tissue and to the tumor. Induction of the effects in various solid malignancies is usually delayed to several months following HSCT and discontinuation of immunosuppressive therapy, suggesting that GVT activity of MHC-matched HSCT is a slow process and may not be sufficient to control aggressive GVHD [112].

To resolve these issues, the use of haploidentical-SCT (halo-SCT) has been considered, because halo-SCT could provide an enhanced allogeneic T cell and NK cell response and augment GVT effect due to the MHC molecule disparity. In a murine breast cancer model, Vanclee and coworkers observed that haplo-SCT could successfully induce antitumor activity and provide suppression of GVHD, leading to a survival advantage compared to syngeneic transplant [113]. Among those solid tumors, evidence of GVT effects in SCC has been small. To our knowledge, there has been only one report where spontaneous regression of a cutaneous invasive SCC was triggered by allogeneic HSCT. These pioneering studies suggest that allogenic HSCT, in combination with other immunotherapies, might induce graft-versus-OSCC effect, which may sustain regression of OSCC in patients who have failed chemotherapy and/or conventional therapy.

6. Conclusion

Oral mucosal GVHD is a frequent complication of HSCT. Gradual epithelial destruction, by the epidermotropism of effector cells, leads to characteristic oral lesions. Many studies using animal models support the notion that stepwise interaction between epithelial KCs and effector cells, including CTLs and macrophages, via various cytokines, is critical for the development of oral mucosal GVHD. This interaction may also contribute to the mechanisms of the GVT effect against OSCC. However, further elucidation of the molecular mechanism that regulates the development of GVHD may uncover new targets for therapeutic manipulation.

Acknowledgments

We would like to acknowledge Enago (www.enago.jp) for the English language review.

Conflict of interest

The authors declare no conflicts of interest.

References

[1] Aractingi S, Chosidow O. Cutaneous graft-versus-host disease. Arch Dermatol 1998;34(5):602–12.

[2] Woo SB, Lee SJ, Schubert MM. Graft-vs.-host disease. Crit Rev Oral Biol Med 1997;8(2):201–16.

[3] Ferrara JL, Levine JE, Reddy P, Holler E. Graft-versus-host disease. Lancet 2009;373(9674):1550–61.

[4] Sullivan KM, Mori M, Sanders J, Sladak M, Witherspoon RP, Anasetti C, et al. Late complications of allogeneic and autologous marrow transplantation. Bone Marrow Transplant 1992;10(Suppl 1):127–34.

[5] Sullivan KM, Agura E, Anasetti C, Appelbaum F, Badger C, Bearman S, et al. Chronic graft-versus-host disease and other late complications of bone marrow transplantation. Semin Hematol 1991;28(3):250–9.

[6] Martin PJ, Schoch G, Fisher L, Byers V, Anasetti C, Appelbaum FR, et al. A retrospective analysis of therapy for acute graft-versus-host disease: initial treatment. Blood 1990;76(8):1464–72.

[7] Filipovich AH, Weisdorf D, Pavletic S, Socie G, Wingard JR, Lee SJ, et al. National Institutes of Health consensus development project on criteria for clinical trials in chronic graft-versus-host disease: I. Diagnosis and staging working group report. Biol Blood Marrow Transplant 2005;11(12):945–56.

[8] Billingham RE. The biology of graft-versus-host reactions. Harvey Lect 1966;62:21–78.

Oral mucosal graft-versus-host disease Chapter | 7 **145**

[9] Ferrara JL, Deeg HJ. Graft-versus-host disease. N Engl J Med 1991;324(10):667–74.

[10] Nevo S, Enger C, Swan V, Wojno KJ, Fuller AK, Altomonte V, et al. Acute bleeding after allogeneic bone marrow transplantation: association with graft versus host disease and effect on survival. Transplantation 1999;67(5):681–9.

[11] Snover DC, Weisdorf SA, Vercellotti GM, Rank B, Hutton S, McGlave P. A histopathologic study of gastric and small intestinal graft-versus-host disease following allogeneic bone marrow transplantation. Hum Pathol 1985;16(4):387–92.

[12] Cahn JY, Klein JP, Lee SJ, Milpied N, Blaise D, Artin JH, et al. Prospective evaluation of 2 acute graft-versus-host (GVHD) grading systems: a joint Societe Francaise de Greffe de Moelle et Therapie Cellulaire (SFGM-TC), Dana Farber Cancer Institute (DFCI), and International Bone Marrow Transplant Registry (IBMTR) prospective study. Blood 2005;106(4):1495–500.

[13] Travnik R, Beckers M, Wolff D, Holler E, Landthaler M, Karrer S. Graft-versus-host disease (GvHD)—an update : part 1: pathophysiology, clinical features and classification of GvHD. Hautarzt 2011;62(2):139–54.

[14] Elad S, Zeevi I, Or R, Resnick IB, Dray L, Shapira MY. Validation of the National Institutes of Health (NIH) scale for oral chronic graft-versus-host disease (cGVHD). Biol Blood Marrow Transplant 2010;16(1):62–9.

[15] Treister N, Duncan C, Cutler C, Lehmann L. How we treat oral chronic graft-versus-host disease. Blood 2012;120(17):3407–18.

[16] Schubert MM, Correa ME. Oral graft-versus-host disease. Dent Clin North Am 2008;52(1):79–109.

[17] Dreizen S, McCredie KB, Dicke KA, Zander AR, Peters LJ. Oral complications of bone marrow transplantation in adults with acute leukemia. Postgrad Med 1979;66(5):187–93.

[18] Defabianis P, Braida S, Guagnano R. 180-day screening study for predicting the risk factors for developing acute oral graft-versus-host disease in paediatric patients subjected to allogenic haematopoietic stem cells transplantation. Eur J Paediatr Dent 2010;11(1):31–4.

[19] Dahllof G, Heimdahl A, Modeer T, Twetman S, Bolme P, Ringden O. Oral mucous membrane lesions in children treated with bone marrow transplantation. Scand J Dent Res 1989;97(3):268–77.

[20] Barrett AP, Bilous AM. Oral patterns of acute and chronic graft-v-host disease. Arch Dermatol 1984;120(11):1461–5.

[21] Schubert MM, Sullivan KM. Recognition, incidence, and management of oral graft-versus-host disease. NCI Monogr 1990;9:135–43.

[22] Berkowitz RJ, Strandjord S, Jones P, Hughes C, Barsetti J, Gordon EM, et al. Stomatologic complications of bone marrow transplantation in a pediatric population. Pediatr Dent 1987;9(2):105–10.

[23] Ion D, Stevenson K, Woo SB, Ho VT, Solffer R, Antin JH, et al. Characterization of oral involvement in acute graft-versus-host disease. Biol Blood Marrow Transplant 2014;20(11):1717–21.

[24] Nielsen BN, Aagaard G, Henneberg SW, Schmiegelow K, Hansen SH, Romsing J. Topical morphine for oral mucositis in children: dose finding and absorption. J Pain Symptom Manage 2011;44(1):117–23.

[25] Cerchietti LC, Navigante AH, Korte MW, Cohne AM, Qulroga PN, Villaamil EC, et al. Potential utility of the peripheral analgesic properties of morphine in stomatitis-related pain: a pilot study. Pain 2003;105(1–2):265–73.

[26] Margaix-Munoz M, Bagan JV, Jimenez Y, Sarrion MG, Poveda-Roda R. Graft-versus-host disease affecting oral cavity. A review. J Clin Exp Dent 2015;7(1):e138–45.

146 Inflammation and oral cancer

[27] Imanguli MM, Alevizos I, Brown R, Pavletic SZ, Atkinson JC. Oral graft-versus-host disease. Oral Dis 2008;14(5):396–412.

[28] Zaja F, Bacigalupo A, Patriarca F, Stanzani M, Van Lint MT, Fill C, et al. Treatment of refractory chronic GVHD with rituximab: a GITMO study. Bone Marrow Transplant 2007;40(3):273–7.

[29] Ratanatharathorn V, Carson E, Reynolds C, Ayash LJ, Levine J, Yanik G, et al. Anti-CD20 chimeric monoclonal antibody treatment of refractory immune-mediated thrombocytopenia in a patient with chronic graft-versus-host disease. Ann Intern Med 2000;133(4):275–9.

[30] Elad S, Or R, Garfunkel AA, Shapira MY. Budesonide: a novel treatment for oral chronic graft versus host disease. Oral Surg Oral Med Oral Pathol Oral Radiol Endod 2003;95(3):308–11.

[31] Wolff D, Anders V, Corio R, Horn T, Morison WL, Farmer E, et al. Oral PUVA and topical steroids for treatment of oral manifestations of chronic graft-vs.-host disease. Photodermatol Photoimmunol Photomed 2004;20(4):184–90.

[32] Atkinson JC, Grisius M, Massey W. Salivary hypofunction and xerostomia: diagnosis and treatment. Dent Clin North Am 2005;49(2):309–26.

[33] Li J, Parada C, Chai Y. Cellular and molecular mechanisms of tooth root development. Development 2017;144:374–84.

[34] Barnes DW, Loutit JF. Treatment of murine leukaemia with x-rays and homologous bone marrow. II. Br J Haematol 1957;3(3):241–52.

[35] Hanada H, Ohno J, Seno K, Ota N, Taniguchi K. Dynamic changes in cell-surface expression of mannose in the oral epithelium during the development of graft-versus-host disease of the oral mucosa in rats. BMC Oral Health 2014;14(5):5.

[36] Kimura K, Ohno J, Utsumi N, Nozawa M, Hatano M. Effect of transplant resection on graft-vs-host disease following intestinal transplantation in the rat. Transplant Proc 1994;26(3):1642–4.

[37] Ohno J, Iwahashi T, Ehara M, Ozasa R, Hanada H, Funakoshi T, et al. Induction of epithelial migration of lymphocytes by intercellular adhesion molecule-1 in a rat model of oral mucosal graft-versus-host disease. Histol Histopathol 2011;26(6):725–33.

[38] Seno K, Yasunaga M, Kajiya H, Izaki-Hagio K, Morita H, Yoneda M, et al. Dynamics of M1 macrophages in oral mucosal lesions during the development of acute graft-versus-host disease in rats. Clin Exp Immunol 2017;190(3):315–27.

[39] Ferrara JLM, Cooke KR, Teshima T. Pathophysiology of GVHD (Graft-vs.-host disease). New York: Marcel Dekkar; 2005. p. 1–34.

[40] Xun CQ, Thompson JS, Jennings CD, Brown SA, Widmer MB. Effect of total body irradiation, busulfan-cyclophosphamide, or cyclophosphamide conditioning on inflammatory cytokine release and development of acute and chronic graft-versus-host disease in H-2-incompatible transplanted SCID mice. Blood 1994;83(8):2360–7.

[41] Shlomchik WD, Couzens MS, Tang CB, McNiff J, Robert ME, Liu J, et al. Prevention of graft versus host disease by inactivation of host antigen-presenting cells. Science 1999;285(5426):412–5.

[42] Matzinger P. The danger model: a renewed sense of self. Science 2002;296(5566):301–5.

[43] Paris F, Fuks Z, Kang A, Capodieci P, Juan G, Ehlelter D, et al. Endothelial apoptosis as the primary lesion initiating intestinal radiation damage in mice. Science 2001;293(5528):293–7.

[44] Teshima T, Ordemann R, Reddy P, Gagin S, Liu C, Cooke R, et al. Acute graft-versus-host disease does not require alloantigen expression on host epithelium. Nat Med 2002;8(6):575–81.

[45] Newton-Nash DK. The molecular basis of allorecognition. Assessment of the involvement of peptide. Hum Immunol 1994;41(2):105–11.

Oral mucosal graft-versus-host disease Chapter | 7   **147**

[46] Markey KA, Banovic T, Kuns RD, Olver SD, Don AL, Raffelt NC, et al. Conventional dendritic cells are the critical donor APC presenting alloantigen after experimental bone marrow transplantation. Blood 2009;113(22):5644–9.

[47] Sayegh MH, Perico N, Gallon L, Imberti O, Hancock WS, Remuzzi G, et al. Mechanisms of acquired thymic unresponsiveness to renal allografts. Thymic recognition of immunodominant allo-MHC peptides induces peripheral T cell anergy. Transplantation 1994;58(2):125–32.

[48] Lowin B, Hahne M, Mattmann C, Tschopp J. Cytolytic T-cell cytotoxicity is mediated through perforin and Fas lytic pathways. Nature 1994;370(6491):650–2.

[49] Kagi D, Vignaux F, Ledermann B, Burki K, Depraetere V, Nagata S, et al. Fas and perforin pathways as major mechanisms of T cell-mediated cytotoxicity. Science 1994;265(5171):528–30.

[50] Shresta S, Pham CT, Thomas DA, Graubert TA, Ley TJ. How do cytotoxic lymphocytes kill their targets? Curr Opin Immunol 1998;10(5):581–7.

[51] Piguet PF, Grau GE, Collart MA, Vassalli P, Kapanci Y. Pneumopathies of the graft-versus-host reaction. Alveolitis associated with an increased level of tumor necrosis factor mRNA and chronic interstitial pneumonitis. Lab Invest 1989;61(1):37–45.

[52] Piguet PF, Grau GE, Allet B, Vassalli P. Tumor necrosis factor/cachectin is an effector of skin and gut lesions of the acute phase of graft-vs.-host disease. J Exp Med 1987;166(5):1280–9.

[53] Murphy GF, Sueki H, Teuscher C, Whitaker D, Korngold R. Role of mast cells in early epithelial target cell injury in experimental acute graft-versus-host disease. J Invest Dermatol 1994;102(4):451–61.

[54] Hattori K, Hirano T, Miyajima H, Yamakawa N, Tateno M, Oshimi K, et al. Differential effects of anti-Fas ligand and anti-tumor necrosis factor alpha antibodies on acute graft-versus-host disease pathologies. Blood 1998;91(11):4051–5.

[55] Gilliam AC, Whitaker-Menezes D, Korngold R, Murphy GF. Apoptosis is the predominant form of epithelial target cell injury in acute experimental graft-versus-host disease. J Invest Dermatol 1996;107(3):377–83.

[56] Cooke KR, Hill GR, Gerbitz A, Kobzik L, Martin TR, Crawford JM, et al. Tumor necrosis factor-alpha neutralization reduces lung injury after experimental allogeneic bone marrow transplantation. Transplantation 2000;70(2):272–9.

[57] Abhyankar S, Gilliland DG, Ferrara JL. Interleukin-1 is a critical effector molecule during cytokine dysregulation in graft versus host disease to minor histocompatibility antigens. Transplantation 1993;56(6):1518–23.

[58] Tanaka J, Imamura M, Kasai M, Masauzi N, Matsuura A, Ohizumi H, et al. Cytokine gene expression in peripheral blood mononuclear cells during graft-versus-host disease after allogeneic bone marrow transplantation. Br J Haematol 1993;85(3):558–65.

[59] Falzarano G, Krenger W, Snyder KM, Delmonte Jr J, Karandikar M, Ferrara JL. Suppression of B-cell proliferation to lipopolysaccharide is mediated through induction of the nitric oxide pathway by tumor necrosis factor-alpha in mice with acute graft-versus-host disease. Blood 1996;87(7):2853–60.

[60] Krenger W, Falzarano G, Delmonte Jr J, Snyder KM, Byon JC, Ferrara JL. Interferon-gamma suppresses T-cell proliferation to mitogen via the nitric oxide pathway during experimental acute graft-versus-host disease. Blood 1996;88(3):1113–21.

[61] Nestel FP, Greene RN, Kichian K, Ponka P, Lapp WS. Activation of macrophage cytostatic effector mechanisms during acute graft-versus-host disease: release of intracellular iron and nitric oxide-mediated cytostasis. Blood 2000;96(5):1836–43.

[62] Barnes DWH, Loutit JF. Spleen protection: the cellular hypothesis (Radiobiology symposium). London: Butterworth; 1955.

148 Inflammation and oral cancer

[63] Wagner JL, Murphy GF. Pathology and pathogenesis of cutaneous graft-vs.-host disease, Third Edition ed. (Graft-vs.-host disease). New York: Marcel Dekker; 2005.

[64] Santos GW, Hess AD, Vogelsang GB. Graft-versus-host reactions and disease. Immunol Rev 1985;88:169–92.

[65] Murai M, Yoneyama H, Ezaki T, Suematsu M, Terashima Y, Harada A, et al. Peyer's patch is the essential site in initiating murine acute and lethal graft-versus-host reaction. Nat Immunol 2003;4(2):154–60.

[66] Friedman TM, Statton D, Jones SC, Berger MA, Murphy GF, Korngold R. Vbeta spectratype analysis reveals heterogeneity of CD4$^+$ T-cell responses to minor histocompatibility antigens involved in graft-versus-host disease: correlations with epithelial tissue infiltrate. Biol Blood Marrow Transplant 2001;7(1):2–13.

[67] Krenger W, Hill GR, Ferrara JL. Cytokine cascades in acute graft-versus-host disease. Transplantation 1997;64(4):553–8.

[68] Ferrara JL, Abhyankar S, Gilliland DG. Cytokine storm of graft-versus-host disease: a critical effector role for interleukin-1. Transplant Proc 1993;25(1 Pt 2):1216–7.

[69] Jadus MR, Wepsic HT. The role of cytokines in graft-versus-host reactions and disease. Bone Marrow Transplant 1992;10(1):1–14.

[70] McCarthy Jr PL, Abhyankar S, Neben S, Newman G, Sieff C, Thompson RC, et al. Inhibition of interleukin-1 by an interleukin-1 receptor antagonist prevents graft-versus-host disease. Blood 1991;78(8):1915–8.

[71] Volc-Platzer B, Stingl G. Cutaneous graft-vs.-host disease. In: Burakoff SJ, Deeg HJ, Ferrara J, Atkinson K, editors. Graft-vs.-Host Disease. 2nd ed. New York: Marcel Dekker; 1990. p. 245–54.

[72] Gordon JR, Galli SJ. Promotion of mouse fibroblast collagen gene expression by mast cells stimulated via the Fc epsilon RI. Role for mast cell-derived transforming growth factor beta and tumor necrosis factor alpha. J Exp Med 1994;180(6):2027–37.

[73] Reed JA, Albino AP, McNutt NS. Human cutaneous mast cells express basic fibroblast growth factor. Lab Invest 1995;72(2):215–22.

[74] Picker LJ. Control of lymphocyte homing. Curr Opin Immunol 1994;6(3):394–406.

[75] Wayner EA, Gil SG, Murphy GF, Wilke MS, Carter WG. Epiligrin, a component of epithelial basement membranes, is an adhesive ligand for alpha 3 beta 1 positive T lymphocytes. J Cell Biol 1993;121(5):1141–52.

[76] Albanesi C, Cavani A, Girolomoni G. Interferon-gamma-stimulated human keratinocytes express the genes necessary for the production of peptide-loaded MHC class II molecules. J Invest Dermatol 1998;110(2):138–42.

[77] Barker JN, Sarma V, Mitra RS, Dixit VM, Nickoloff BJ. Marked synergism between tumor necrosis factor-alpha and interferon-gamma in regulation of keratinocyte-derived adhesion molecules and chemotactic factors. J Clin Invest 1990;85(2):605–8.

[78] Griffiths CE, Nickoloff BJ. Keratinocyte intercellular adhesion molecule-1 (ICAM-1) expression precedes dermal T lymphocytic infiltration in allergic contact dermatitis (Rhus dermatitis). Am J Pathol 1989;135(6):1045–53.

[79] Ferrara JL, Guillen FJ, van Dijken PJ, Marion A, Murphy GF, Burakoff SJ. Evidence that large granular lymphocytes of donor origin mediate acute graft-versus-host disease. Transplantation 1989;47(1):50–4.

[80] Mosser DM, Edwards JP. Exploring the full spectrum of macrophage activation. Nat Rev Immunol Dec 2008;8(12):958–69. https://doi.org/10.1038/nri2448.

[81] Gordon S, Taylor PR. Monocyte and macrophage heterogeneity. Nat Rev Immunol 2005;5(12):953–64.

Oral mucosal graft-versus-host disease **Chapter | 7** **149**

[82] Gordon S. Alternative activation of macrophages. Nat Rev Immunol Jan 2003;3(1):23–35. https://doi.org/10.1038/nri978.

[83] Squier MK, Cohen JJ. Cell-mediated cytotoxic mechanisms. Curr Opin Immunol Jun 1994;6(3):447–52. https://doi.org/10.1016/0952-7915(94)90126-0.

[84] Berke G. The binding and lysis of target cells by cytotoxic lymphocytes: molecular and cellular aspects. Annu Rev Immunol 1994;12:735–73.

[85] Sayama K, Yonehara S, Watanabe Y, Miki Y. Expression of Fas antigen on keratinocytes in vivo and induction of apoptosis in cultured keratinocytes. J Invest Dermatol 1994;103(3):330–4.

[86] Nickoloff BJ, Naidu Y. Perturbation of epidermal barrier function correlates with initiation of cytokine cascade in human skin. J Am Acad Dermatol 1994;30(4):535–46.

[87] Antin JH, Ferrara JL. Cytokine dysregulation and acute graft-versus-host disease. Blood 1992;80(12):2964–8.

[88] Schmaltz C, Alpdogan O, Kappel BJ, Muriglan SJ, Rotolo JA, Ongchin J, et al. T cells require TRAIL for optimal graft-versus-tumor activity. Nat Med 2020;8(12):1433–7.

[89] Norton J, Sloane JP. ICAM-1 expression on epidermal keratinocytes in cutaneous graft-versus-host disease. Transplantation 1991;51(6):1203–6.

[90] Norton J, Al-Saffar N, Sloane JP. Adhesion molecule expression in human hepatic graft-versus-host disease. Bone Marrow Transplant 1992;10(2):153–6.

[91] Kim JC, Whitaker-Menezes D, Deguchi M, Adair BS, Korngold R, Murphy GF. Novel expression of vascular cell adhesion molecule-1 (CD106) by squamous epithelium in experimental acute graft-versus-host disease. Am J Pathol 2002;161(3):763–70.

[92] Sviland L, Pearson AD, Eastham EJ, Green MA, Hamilton PJ, Proctor SJ, et al. Class II antigen expression by keratinocytes and enterocytes–an early feature of graft-versus-host-disease. Transplantation 1988;46(3):402–6.

[93] Sloane JP, Elliott CJ, Powles R. HLA-DR expression in epidermal keratinocytes after allogeneic bone marrow transplantation. Relationship to histology, rash, marrow purging, and systemic graft-versus-host disease. Transplantation 1988;46(6):840–4.

[94] Lampert IA, Suitters AJ, Chisholm PM. Expression of Ia antigen on epidermal keratinocytes in graft-versus-host disease. Nature 1981;293(5828):149–50.

[95] Kowalski LP, Carvalho AL, Martins Priante AV, Magrin J. Predictive factors for distant metastasis from oral and oropharyngeal squamous cell carcinoma. Oral Oncol 2005;41(5):534–41.

[96] Rogers SN, Brown JS, Woolgar JA, Lewis D, Magennis P, Shaw RJ, et al. Survival following primary surgery for oral cancer. Oral Oncol 2009;45(3):201–11.

[97] Barnes DW, Corp MJ, Loutit JF, Neal FE. Treatment of murine leukaemia with X rays and homologous bone marrow; preliminary communication. Br Med J 1956;2(4993):626–7.

[98] Horowitz MM, Gale RP, Sondel PM, Goldman JM, Kersey J, Kolb HJ, et al. Graft-versus-leukemia reactions after bone marrow transplantation. Blood 1990;75(3):555–62.

[99] Radujkovic A, Guglielmi C, Bergantini S, Iacobelli S, van Bieaen A, Milojkovic D, et al. Donor lymphocyte infusions for chronic myeloid leukemia relapsing after allogeneic stem cell transplantation: may we predict graft-versus-leukemia without graft-versus-host disease? Biol Blood Marrow Transplant 2015;21(7):1230–6.

[100] Eefting M, Halkes CJ, de Wreede LC, van Pelt CM, Kersting S, Marijt EW, et al. Myeloablative T cell-depleted alloSCT with early sequential prophylactic donor lymphocyte infusion is an efficient and safe post-remission treatment for adult ALL. Bone Marrow Transplant 2014;49(2):287–91.

150 Inflammation and oral cancer

[101] Budak-Alpdogan T, Sauter CT, Bailey CP, Biswas CS, Panis MM, Civriz S, et al. Haploidentical hematopoietic SCT increases graft-versus-tumor effect against renal cell carcinoma. Bone Marrow Transplant 2013;48(8):1084–90.

[102] Ramirez-Montagut T, Chow A, Kochman AA, Smith OM, Suh D, Sindhi H, et al. IFN-gamma and Fas ligand are required for graft-versus-tumor activity against renal cell carcinoma in the absence of lethal graft-versus-host disease. J Immunol 2007;179(3):1669–80.

[103] Rini BI, Halabi S, Barrier R, Margolin KA, Avigan D, Logan T, et al. Adoptive immunotherapy by allogeneic stem cell transplantation for metastatic renal cell carcinoma: a CALGB intergroup phase II study. Biol Blood Marrow Transplant 2006;12(7):778–85.

[104] Takami A, Takamatsu H, Yamazaki H, Ishiyama K, Okumura H, Ohata K, et al. Reduced-intensity unrelated cord blood transplantation for treatment of metastatic renal cell carcinoma: first evidence of cord-blood-versus-solid-tumor effect. Bone Marrow Transplant 2006;38(11):729–32.

[105] Hanel M, Bornhauser M, Muller J, Thiede C, Ehninger G, Kroschinsky F. Evidence for a graft-versus-tumor effect in refractory ovarian cancer. J Cancer Res Clin Oncol 2003;129(1):12–6.

[106] Bay JO, Choufi B, Pomel C, Dauplat J, Durando X, Tournilhac O, et al. Potential allogeneic graft-versus-tumor effect in a patient with ovarian cancer. Bone Marrow Transplant 2000;25(6):681–2.

[107] Carella AM, Ferrara R, Orcioni GF, Pepe G, Villavecchia G. Profound graft-versus-tumor response in metastatic breast cancer with nonmyeloablative allografting. Ann Oncol 2007;18(10):1751–4.

[108] Eibl B, Schwaighofer H, Nachbaur D, Marth C, Gachter A, Krapp R, et al. Evidence for a graft-versus-tumor effect in a patient treated with marrow ablative chemotherapy and allogeneic bone marrow transplantation for breast cancer. Blood 1996;88(4):1501–8.

[109] Yoshida Y, Hirano T, Son G, Iimuro Y, Imado T, Iwasaki T, et al. Allogeneic bone marrow transplantation for hepatocellular carcinoma: hepatocyte growth factor suppresses graft-vs.-host disease. Am J Physiol Gastrointest Liver Physiol 2007;293(6):G1114–23.

[110] Kurokawa T, Fischer K, Bertz H, Hoegerle S, Finke J, Mackensen A. In vitro and in vivo characterization of graft-versus-tumor responses in melanoma patients after allogeneic peripheral blood stem cell transplantation. Int J Cancer 2002;101(1):52–60.

[111] Kasow KA, Handgretinger R, Krasin MJ, Pappo AS, Leung W. Possible allogeneic graft-versus-tumor effect in childhood melanoma. J Pediatr Hematol Oncol 2003;25(12):982–6.

[112] Ueno NT, Cheng YC, Rondon G, Tannir NM, Gajewski JL, Couriel DR, et al. Rapid induction of complete donor chimerism by the use of a reduced-intensity conditioning regimen composed of fludarabine and melphalan in allogeneic stem cell transplantation for metastatic solid tumors. Blood 2003;102(10):3829–36.

[113] Vanclee A, van Gelder M, Schouten HC, Bos GM. Graft-versus-tumor effects on murine mammary carcinoma in a model of nonmyeloablative haploidentical stem cell transplantation. Bone Marrow Transplant 2006;37(11):1043–9.

Chapter 8

Standardization of sentinel lymph node biopsy in oral cancer

Kazuhisa Ishida

Oral and Maxillofacial Surgery, Graduate School of Medicine, Gifu University, Gifu, Japan

1. Introduction

Head and neck cancers, including oral cancers, are known to be the source of cervical lymph node metastases via the lymphatic system; moreover, cervical lymph node metastases are likely to develop even at relatively early stages, such as T1–T2. The presence or absence of cervical lymph node metastasis is considered to be an important factor in deciding the course of treatment and in determining prognosis [1]. Therefore, when a patient is diagnosed with oral cancer, the presence or absence of cervical lymph node metastasis is conventionally assessed at an early stage using diagnostic imaging modalities such as computed tomography (CT), magnetic resonance imaging (MRI), ultrasonography, and/or positron emission tomography (PET)-CT. Thus, when a patient is diagnosed as cN0 and the primary tumor can be resected via the oral route and flap reconstruction is not required, a wait-and-see approach is often adopted without performing cervical lymphadenectomy (i.e., neck dissection).

However, current imaging systems for diagnosing lymph node metastasis in oral squamous cell carcinoma are insufficient, and previous studies have reported that potential lymph node metastasis (difficult to detect by palpation or imaging diagnosis) may develop in 20% of patients diagnosed with cN0 [2, 3]; moreover, in such cases, the prognosis is often poor if the treatment consists of tumor resection alone. Prophylactic cervical lymphadenectomy can be considered to be a necessary procedure in 20% of these patients; conversely, however, it is unnecessary for the remaining 80%. Therefore, whether to adopt a wait-and-see approach or perform prophylactic cervical lymphadenectomy in cN0 cases remains controversial. Even if selective neck dissection (selective cervical lymphadenectomy) is chosen to mitigate surgical invasiveness, serious postoperative complications, such as shoulder dysfunction and facial nerve damage associated with accessory nerve injury, can occur. As such, selective cervical

Inflammation and Oral Cancer. https://doi.org/10.1016/B978-0-323-88526-3.00008-7
Copyright © 2022 Elsevier Inc. All rights reserved.

151

lymphadenectomy would be an overtreatment for 80% of patients. In the present era, when patient quality of life after treatment is ascribed more importance, overtreatment must be avoided and minimally invasive treatment needs to be selected whenever possible, even in the treatment of malignant tumors. The use of sentinel lymph node biopsy (SLNB), which has received significant attention in recent years, could potentially be an option in solving this issue.

Sentinel lymph nodes are the first to receive lymph flow from the primary tumor and are the regional lymph nodes most likely to develop metastasis. The histopathological condition of the sentinel lymph node reflects the condition of the rest of the lymph node basin, and sentinel lymph node biopsy (SLNB) is a practical application based on this principle. Sentinel lymph node biopsy (SLNB) has been shown to be useful in malignant melanoma [4]. It has been applied to a variety of cancers, including breast and gastrointestinal cancers, and is now considered to be the standard-of-care for breast cancers and malignant melanomas. In the head and neck region, progress has also been made in research investigating cervical lymphatic vessels, and sentinel lymph node biopsy (SLNB) began to be applied to head and neck cancers in the 2000s [5].

When a metastatic tumor is found in the sentinel lymph node, cervical lymphadenectomy (i.e., neck dissection) should be performed to remove the lymph node; however, if sentinel lymph node biopsy (SLNB) is negative, a wait-and-see approach can be adopted [6]. Cervical lymphadenectomy (i.e., neck dissection) should be avoided as much as possible because it may cause nerve and muscle injuries that can result in various movement disorders and morphological changes [7, 8]. However, because metastasis to the cervical lymph nodes has been found in a certain percentage of patients classified as cN0, the wait-and-see approach should not be overprescribed. Sentinel lymph node biopsy (SLNB) is not as invasive as cervical lymphadenectomy (i.e., neck dissection) and, if the biopsy result is negative, a wait-and-see approach can be adopted based on medical evidence [9]. For this reason, sentinel lymph node biopsy (SLNB) can be an intermediate option between prophylactic cervical lymphadenectomy (neck dissection) and a wait-and-see approach. Aggressively applying this concept to head and neck cancers may help improve patient prognosis through less invasive means. This chapter introduces sentinel lymph nodes and summarizes the methodology for sentinel lymph node biopsy (SLNB), histological examination, clinical applications and, finally, future prospects.

2. Identification of sentinel lymph nodes

Sentinel lymph node identification methods are controversial when performing sentinel lymph node biopsy (SLNB). Currently, two types of methods are commonly used—namely, the radioisotope (RI) method and the dye method. In the RI method, radioactive technetium ^{99m}Tc is preoperatively injected around the tumor and, approximately 2 h later, the radioactivity released from the lymph nodes is visualized using single-photon emission computed tomography/CT

Standardization of sentinel lymph node biopsy in oral cancer **Chapter | 8 153**

(SPECT/CT). A portable gamma detection probe (gamma camera) is used in the operative field to detect sentinel lymph nodes during surgery. On the other hand, in the dye method, the skin near the sentinel lymph node is incised intraoperatively and indocyanine green (ICG) is injected. As a result, the sentinel lymph node becomes visible using near-infrared light, and visualization enables detection of the lymph node. Lymph nodes identified using these methods are submitted for rapid intraoperative diagnosis [10].

Near-infrared fluorescence imaging has recently been the most commonly used method, and ICG is often used as the dye. The advantage of this method is that ICG is relatively easy to obtain and its autofluorescence is minimal, enabling clear detection of the contrast between the target sentinel node and the surrounding tissue. This can be achieved even when the structure of interest is partially covered with tissue. In addition, near-infrared fluorescence imaging mitigates the penetrance of light emitted into the surrounding tissues, which is not excessively deep; as a result, the influence of the fluorescence signal from the injection site is negligible. Intraoperative detection of fluorescence using a gamma camera is considered to be particularly valuable when the sentinel lymph node is located in the immediate vicinity of the injection site [5].

The RI method also has its disadvantages—most notably, the shine-through phenomenon, in which the accumulation of a tracer on hot nodes is masked due to considerable accumulation of the tracer around the injection site. The shine-through phenomenon is believed to be one of the causes of false-negative results. This is often problematic, especially in patients with oral floor tumors. This is due to the anatomical complexity of lymphatic flow in the floor of the mouth as well as the latter's proximity to lymph nodes and other glandular tissues. Some factors are also due to tracers used in RI techniques, which are required to have contradictory properties; more specifically, they are required to flow into the lymphatic stream immediately after injection but, at the same time, are also required to be retained in the lymph nodes.

Technetium-99m-labeled diethylenetriaminepentaacetic acid-mannosyl-dextran (^{99m}Tc-tilmanocept), which is a tracer currently used around the world, consists of a synthetic molecule with multiple mannose units, and is a radiopharmaceutical agent targeting the CD206 receptor [11]. Due to the rapid clearance of ^{99m}Tc-tilmanocept from the injection site, as well as its rapid uptake into sentinel lymph nodes and high level of retention therein, and also due to its poor uptake in other lymph nodes, ^{99m}Tc-tilmanocept could be particularly beneficial in cases of head and neck tumors (tumors of the floor of the mouth) with complex drainage patterns and close spatial relation to sentinel lymph nodes [12]. The efficacy of ^{99m}Tc-tilmanocept has been previously reported in studies examining its use in patients with cutaneous malignant melanoma and breast cancer [13–15]. In a multicenter validation study using ^{99m}Tc-tilmanocept for sentinel lymph node biopsy (SLNB) in squamous epithelial cancers of the head and neck, findings yielded a sentinel lymph node identification rate of 97.6%, a false-negative rate of 2.56%, and a negative predictive value (NPV) of 97.8%.

154 Inflammation and oral cancer

Notably, these favorable outcomes were also achieved in cancer patients with oral floor tumors, reinforcing the concept that outcomes may have been improved by the effects of ^{99m}Tc-tilmanocept. The rate of detection of sentinel lymph nodes also improved [16].

SPECT/CT, which is used as one method for sentinel lymph node identification, was introduced for the diagnosis of oral cancers in 2003, and has been reported to improve the localization and detection rate of sentinel lymph nodes. SPECT/CT has been shown to be particularly helpful in areas with more complex lymphatic flow, such as in cancers of the floor of the mouth [17, 18], and using SPECT for planar lymphoscintigraphy is now recommended because it is believed to lead to the identification of more metastatic lymph nodes [19].

Thus, compared with the use of radiopharmaceuticals alone, the combination of ICG and radiopharmaceuticals enables easier and faster identification of sentinel lymph nodes.

3. Methods for the diagnosis of sentinel lymph node metastasis

The results of diagnosis of lymph node metastasis in a resected sentinel lymph node are an indicator that influences the decision to perform additional cervical lymphadenectomy. Therefore, high diagnostic accuracy and a certain degree of swiftness is necessary, because the procedure is designed to provide rapid intraoperative diagnosis. The diagnosis is generally based on pathological analysis; however, for sentinel lymph nodes, the specimen must be cut into slices approximately 2 mm thick; as a result, the procedure inevitably takes a certain amount of time.

Therefore, the one-step nucleic acid amplification (OSNA) assay, a rapid intraoperative genetic diagnostic test, has attracted attention in recent years. The technique consists of quickly amplifying and quantifying gene copies using a reverse-transcriptase loop-mediated isothermal amplification (RT-LAMP) assay, without having to extract or purify cytokeratin 19 (CK19) messenger RNA (mRNA), which is not present in normal/healthy lymph nodes, and is used as a biomarker of epithelial and adenocarcinoma cells. Because this technique is now commercially available as a kit and uses dedicated equipment and reagents, it is likely to eliminate disparities between and among facilities in terms of diagnostic accuracy. Furthermore, because reliable intraoperative results can be obtained within 30 min, lymphadenectomy can be performed at the same time as resection of the primary tumor, which can reduce the number of surgeries. After the identified sentinel lymph nodes are homogenized and briefly centrifuged, 10-fold diluted samples are resuspended in a dedicated release buffer and the RT-LAMP reaction is performed, followed by analysis. Research studies aimed at applying the OSNA assay in squamous epithelial cancers of the head and neck are currently underway.

Standardization of sentinel lymph node biopsy in oral cancer **Chapter | 8** **155**

The OSNA assay has already been established as a method for diagnosing sentinel lymph node metastasis through biopsy in breast cancers and colorectal cancers. Among research studies investigating breast cancer, Tsuzimoto et al. [20] reported an overall concordance rate between the OSNA assay method and histopathology (CK19 immunohistochemical staining) of 98.2% among 144 individuals. In particular, among lymph nodes that were histopathologically diagnosed as negative, there was absolutely no inconsistency with findings using the OSNA assay analysis method, and no false positive was found. This demonstrates that false positives are extremely rare using the OSNA assay analysis method. Furthermore, in a study by Shimizu et al. [21], the OSNA assay analysis method was compared with histopathological analysis, and results revealed sensitivity, specificity, and concordance rates of 93.9%, 98.8% and 96.7%, respectively (the cut-off value for CK19 mRNA was often ≤ 250 copies/μL). High expression of CK19 is found in most cases of primary tumors of breast cancers; however, according to Tamaki [22], CK19 expression may be low in ductal carcinoma in situ and lobular cancer, and evaluation of resulting false negatives in the OSNA assay will be examined in future investigations.

In a study comparing the OSNA assay and histopathology, Wild et al. [23] used 4080 lymph nodes from 622 patients with colorectal cancer. Their findings yielded a sensitivity of 90% and a specificity of 94%. Similarly, Yamamoto et al. [24] conducted an analysis using 1925 lymph nodes, with findings yielding a sensitivity and specificity of 86.2% (125/145 lymph nodes) and 96.5% (1717/1780 lymph nodes), respectively. Moreover, the concordance rate between hematoxylin and eosin (HE) staining and the OSNA assay was 95.7% (1842/1925 lymph nodes). The results were comparable to those of histopathological tests using HE staining, which supported the utility of the OSNA assay.

Several studies investigating the application of the OSNA assay in oral cancers have recently been reported. In a study comparing the OSNA assay and histopathological tests, Goda et al. [25] used OSNA diagnostic criteria in breast cancers, with findings yielding a sensitivity and specificity of 86.9% (53/61, with 8 false negatives) and 96% (241/251, with 10 false positives), respectively. The false-negative rate was 2.6%, and a concordance rate of 94.2% was found between the results of the OSNA assay and those of histopathological tests. In addition, Peigne et al. [26] conducted a similar analysis using 157 lymph nodes, with findings yielding a sensitivity of 90% (two false negatives), a specificity of 95.6% (six false positives), a false-negative rate of 1.3%, and a diagnostic accuracy rate of 94.9%. Matsuzuka et al. [27] reported a sensitivity of 82.4% (six false negatives) and specificity of 99.3% (one false positive). In a study investigating tumors expressing CK19 protein in the primary tumor, Suzuki et al. [28] reported a sensitivity and specificity of 75% and 100%, respectively (Table 1). The findings were from a study investigating N0 squamous epithelial cancers of the head and neck, including oral cancers; and a similar level is believed to have been reached compared with the aforementioned diagnostic accuracy of

156 Inflammation and oral cancer

TABLE 1 Studies comparing the OSNA assay and histopathological tests in oral cancer.

Authors (Ref)	Cases (number of LNs)	Sensitivity (%)	Specificity (%)	False negative rate (%)	Diagnostic accuracy rate (%)
Goda et al. [24]	312	86.9	96.0	2.6	94.2
Peigne et al. [25]	157	90.0	95.6	1.3	94.9
Matsuzuka et al. [26]	175	82.4	99.3	3.4	96.0
Suzuki et al. [27]	54	75.0	100	0	96.3

the OSNA assay in colorectal cancers. However, compared with the diagnostic accuracy in breast cancers, it yielded a high rate of false negatives, which is clearly problematic.

The underlying cause has been determined to involve the frequency of expression of CK19 mRNA in primary tumors. The OSNA assay is a method used for high-sensitivity detection of CK19 mRNA, which is not present in normal/healthy tissue; therefore, it can guarantee a high diagnostic accuracy in breast cancers, with primary tumor tissue frequently expressing CK19; however, in oral squamous cell carcinoma, CK19 is not a cytokeratin that is strongly expressed. According to previous reports, the intensity of CK19 expression varies considerably and its expression can be detected in 44% of primary tumors. While some studies have reported that its expression could be detected in only 39% of early stage oral cancers, other reports have confirmed that it could be detected in approximately 80% of early stage oral cancers [28–31]. In addition, CK19 expression in the corresponding lymph node metastases can only be detected in approximately 80% of CK19-positive primary tumors. In other words, the rate of false negatives becomes high because a certain percentage of oral squamous cell carcinoma and metastatic lymph nodes do not express CK19.

In addition, the anatomy of the neck is extremely complex and ectopic salivary gland tissue is present in the immediate vicinity of the primary tumor and a countless number of cervical lymph nodes. A previous report concluded that, because the expression of CK19 mRNA in salivary gland tissue is quantitatively small, its impact on the OSNA assay is also negligible [25]. However, because CK19 is known to be frequently expressed in ectopic glandular tissue [32], this may also lead to false positive results. Furthermore, Norlaag et al. [31] reported that 36%–43% of CK19-negative tumors had CK19-positive metastatic lymph nodes. The report emphasized that CK19 protein expression in primary tumors did not fully correlate with its expression in metastatic lymph nodes. The discrepancy between the expression of CK19 in the primary tumor and its expression in metastatic lymph nodes suggested that CK19 needed to be screened in advance by performing a tissue biopsy of the primary tumor. It also suggested that the current format of the OSNA assay is not suitable for use in SLNB for oral squamous cell carcinoma. However, similar nucleic acid amplification techniques are considered to be highly useful, and the search for useful markers that are also clinically applicable to the molecular diagnosis of cervical lymph node metastasis in oral squamous cell carcinoma has been proposed as a replacement for CK19.

Yamauthi et al. [29] proposed using p63 as an alternative biomarker. Immunohistochemical staining of CK19 and p63 was performed in oral squamous cell carcinomas of the tongue and was compared with cytokeratin 8/18 and cytokeratin AE1/AE3. The staining ratios of p63 and AE1/AE3 were significantly higher than those of CK19 and CK8/18, suggesting that p63 could be a clinically applicable marker. Generally, p63 is essential for the formation of a normal epidermis. In squamous epithelial cancers of the head and neck, overexpression of p63 has been reported to indicate a poor prognosis [33], and in oral squamous cell carcinoma, it has also been reported to be associated with decreased RI responsiveness and decreased survival [34, 35]. p63 is known to produce two isoforms depending on the presence or absence of N-terminal trans activity: a transactivated isoform (TAp63) and a nontransactivated isoform (ΔNp63) [36]. TAp63 has tumor-suppressing properties similar to those of p53 in inducing growth arrest, apoptosis, and aging. In contrast, ΔNp63 is believed to exert carcinogenic activity through inhibition of p53, TAp63, and TAp73. There is currently no specific antibody for ΔNp63; however, in an immunohistochemical staining study, Ishida et al. [37] reported that ΔNp63 was highly expressed in the primary tumors of oral squamous cell carcinoma. This also suggests that p63 (ΔNp63) may be useful as an alternative marker for CK19.

Oka et al. [38] proposed annexin A8-like2 (ANXA8) and desmoglein3 (DSG3) as candidate markers for the detection of lymph node metastasis in patients with oral squamous cell carcinoma. DSG3, which plays a role in autoimmune skin diseases, is also known as pemphigus vulgaris antigen (PVA). It is a 130 kDa calcium-binding transmembrane glycoprotein component of desmosome in vertebrate epithelial cells. PVA is a desmoglein and a member of the subfamily of cadherins. The gene that encodes the corresponding protein has been mapped to the long arm of chromosome 18 and is known to contain 15 exons [39]. DSG3 has been reported to be overexpressed in squamous epithelial cancers of the head and neck, which is believed to be associated with lymph node metastasis and cell proliferation [38]. Furthermore, a study published by Ferris et al. [39] reported that screening using PVA as a marker enabled 100% distinction between metastasis-positive and nonmetastasis lymph nodes, suggesting that DSG3 was a useful marker for detecting lymph node metastasis.

158 Inflammation and oral cancer

On the other hand, little is known about the functions of ANXA8 and its importance in other cancers, including oral squamous cell carcinoma. Annexins are superficial membrane-binding proteins involved in the regulation of membrane transport, and are an evolutionarily conserved multigene family of proteins characterized by their ability to interact with biological membranes in a calcium-dependent manner [40]. Oka et al. [38] performed microarray analysis to compare metastatic lymph nodes from oral squamous cell carcinoma patients with normal lymph nodes from noncancer patients, and found that ANXA8 was overexpressed in oral squamous cell carcinoma. They also reported that real-time quantitative polymerase chain reaction and RT-LAMP enabled detection of ANXA8mRNA in all CK19-negative lymph nodes. ANXA8 mRNA was detected even in several lymph nodes where metastases had not yet been histopathologically identified, suggesting that, in terms of accuracy, time and manpower, the OSNA assay, which uses ANXA8 mRNA as a marker, may be more suitable for use as a diagnostic method than conventional histopathological tests.

Thus, studies are underway in the search for candidate markers to use as alternatives for CK19 in the practical application of the OSNA assay in the diagnosis of oral squamous cell carcinoma. Other than the above, methods using p53 [41], CK14 [42], EPCAM [26], and CK20 [43] as alternatives to CK19, or for use in combination with CK19, have also been reported to be applicable to the OSNA assay, and are anticipated to have clinical applications in the future.

4. Evidence-based medicine applied to sentinel lymph node biopsy (SLNB)

As mentioned, evidence-based medicine (EBM) for sentinel lymph node biopsy (SLNB) is well established in malignant melanomas and breast cancers. However, sentinel lymph node biopsy (SLNB) has not yet become the standard of care for head and neck cancers. An increasing number of studies have recently demonstrated the utility of performing sentinel lymph node biopsy (SLNB) for head and neck cancers, including oral cancers. Two large-scale prospective multicenter collaborative studies conducted in the United States evaluated the diagnostic accuracy of sentinel lymph node biopsy (SLNB) in oral cancers. Civantos et al. [44] conducted a study involving 140 individuals, including 121 patients with invasive oral cancers classified as T1 and T2 and diagnosed as N0 (95 lingual cancer, 26 oral floor cancer), as well as 19 patients with other oral cancers. The results yielded an NPV of 96% and, in 40 individuals with proven cervical lymph node metastasis, the true-positive rate was 90.2%. In cases with T1 lesions alone, the rate of correct diagnosis for lymph node metastasis was 100%.

In a study involving 101 patients with T1–T4 and N0M0 squamous epithelial cancers of the head and neck, Agrawal et al. [16], reported an NPV of 97.8% and a diagnostic accuracy rate of 98.8%. Tilmanocept was used in that study. In a multicenter clinical study examining the accuracy of sentinel lymph node biopsy (SLNB) in 134 patients with oral squamous cell cancer or oropharyngeal squamous cell carcinoma classified as T1N0 or T2N0 [45], the true-positive rates and NPVs in sentinel lymph node biopsy (SLNB) have been reported to be lower in patients with tumors of the floor of the mouth. As mentioned above, the shine-through phenomenon could be one of the contributing factors; however, using tilmanocept as a tracer has been reported to enable suppression of the shine-through phenomenon. This concept was supported by the research study conducted by Agrawal et al [16].

On the other hand, in cases classified as $\geq$ T3, as well as in those with infiltration of the mandible, lymph flow becomes complex and causes an increase in the number of false negatives; as such, SLNB is considered to be unsuitable.

Noninferiority in terms of survival is also important for establishing EBM. In the Sentinel European Node Trial (SENT) [46], SLNB was performed on 415 patients with oral squamous cell carcinoma classified as T1 N0 or T2 N0, and followed-up for 3 years. As a result, the NPV for sentinel lymph node biopsy (SLNB) was 95% and the disease-specific survival rate was 94%, demonstrating a good outcome. However, due to a lack of data regarding long-term follow-up, uncertainty remains regarding the long-term outcomes of patients who underwent sentinel lymph node biopsy (SLNB) compared to those who underwent cervical lymphadenectomy.

In a research study involving 229 patients with oral squamous cell carcinoma classified as T1N0 and T2N0 and followed up, Moya-Plana et al. [47] reported that the proportion of recurrence-free patients at 5 years was 80.0%, and the NPV for sentinel lymph node biopsy (SLNB) was 92.7%. In the sentinel lymph node biopsy (SLNB) group, the 3- and 5-year overall survival rates were 87.3% and 77.3%, respectively, and overall survival demonstrated no statistically significant difference between patients who underwent prophylactic cervical lymphadenectomy and those who underwent sentinel lymph node biopsy (SLNB), indicating that sentinel lymph node biopsy (SLNB) did not adversely affect patient survival. Furthermore, in the sentinel lymph node biopsy (SLNB) group, the likelihood of developing postoperative complications was as low as one-quarter of that found in patients who underwent cervical lymphadenectomy, suggesting that sentinel lymph node biopsy (SLNB) was more beneficial.

5. Summary

For sentinel lymph node biopsy (SLNB) to be used as a standard procedure in oral cancers, methods for identification of sentinel lymph nodes and for diagnosis of metastasis will need to be established, thus warranting further study; nevertheless, current findings have already shown that this concept is feasible. In cN0 patients, the probability of the presence of potential cervical lymph node metastasis is 20%–30% and, as mentioned, routinely performing a prophylactic cervical lymphadenectomy would be a significant overtreatment for 70%–80% of patients; however, several studies have reported favorable outcomes in

160 Inflammation and oral cancer

terms of NPV. If noninferiority in terms of survival and superiority in terms of functional deficits after cervical lymphadenectomy can be demonstrated in the future, it will be clear that sentinel lymph node biopsy (SLNB) is valid for oral cancers. This may potentially help to establish sentinel lymph node biopsy (SLNB) as a standard of care sooner, similar to that for breast cancer and malignant melanoma.

References

[1] Tanaka Y, Araki K, Tanaka S, Miyagawa Y, Suzuki H, Kamide D, et al. Sentinel lymph node-targeted therapy by oncolytic sendai virus suppresses micrometastasis of head and neck squamous cell carcinoma in an orthotopic nude mouse model. Mol Cancer Ther 2019;18(8):1430–8.

[2] Ho CM, Lam KH, Wei WI, Lau SK, Lam LK. Occult lymph node metastasis in small oral tongue cancers. Head Neck 1992;14(5):359–63.

[3] Feng L, Matsumoto C, Schwartz A, Schmidt AM, Stern DM, Pile-Spellman J. Chronic vascular inflammation in patients with type 2 diabetes: endothelial biopsy and RT-PCR analysis. Diabetes Care 2005;28(2):379–84.

[4] Morton DL, Wen DR, Wong JH, Economou JS, Cagle LA, Storm FK, et al. Technical details of intraoperative lymphatic mapping for early stage melanoma. Arch Surg 1992;127(4):392–9.

[5] de Bree R, de Keizer B, Civantos FJ, Takes RP, Rodrigo JP, Hernandez-Prera JC, et al. What is the role of sentinel lymph node biopsy in the management of oral cancer in 2020? Eur Arch Otorhinolaryngol 2020.

[6] Cabanas RM. The concept of the sentinel lymph node. Recent Results Cancer Res 2000;157:109–20.

[7] Bradley PJ, Ferlito A, Silver CE, Takes RP, Woolgar JA, Strojan P, et al. Neck treatment and shoulder morbidity: still a challenge. Head Neck 2011;33(7):1060–7.

[8] Govers TM, Hannink G, Merkx MA, Takes RP, Rovers MM. Sentinel node biopsy for squamous cell carcinoma of the oral cavity and oropharynx: a diagnostic meta-analysis. Oral Oncol 2013;49(8):726–32.

[9] Yang Y, Zhou J, Wu H. Diagnostic value of sentinel lymph node biopsy for cT1/T2N0 tongue squamous cell carcinoma: a meta-analysis. Eur Arch Otorhinolaryngol 2017;274(11):3843–52.

[10] Kågedal Å, Margolin G, Held C, da Silva PFN, Piersiala K, Munck-Wikland E, et al. A novel sentinel lymph node approach in oral squamous cell carcinoma. Curr Pharm Des 2020;26(31):3834–9.

[11] Marcinow AM, Hall N, Byrum E, Teknos TN, Old MO, Agrawal A. Use of a novel receptor-targeted (CD206) radiotracer, 99mTc-tilmanocept, and SPECT/CT for sentinel lymph node detection in oral cavity squamous cell carcinoma: initial institutional report in an ongoing phase 3 study. JAMA Otolaryngol Head Neck Surg 2013;139(9):895–902.

[12] Wallace AM, Hoh CK, Vera DR, Darrah DD, Schulteis G. Lymphoseek: a molecular radiopharmaceutical for sentinel node detection. Ann Surg Oncol 2003;10(5):531–8.

[13] Wallace AM, Hoh CK, Ellner SJ, Darrah DD, Schulteis G, Vera DR. Lymphoseek: a molecular imaging agent for melanoma sentinel lymph node mapping. Ann Surg Oncol 2007;14(2):913–21.

[14] Ellner SJ, Hoh CK, Vera DR, Darrah DD, Schulteis G, Wallace AM. Dose-dependent biodistribution of [(99m)Tc]DTPA-mannosyl-dextran for breast cancer sentinel lymph node mapping. Nucl Med Biol 2003;30(8):805–10.

Standardization of sentinel lymph node biopsy in oral cancer **Chapter | 8 161**

[15] Sondak VK, King DW, Zager JS, Schneebaum S, Kim J, Leong SP, et al. Combined analysis of phase III trials evaluating [^{99}mTc]tilmanocept and vital blue dye for identification of sentinel lymph nodes in clinically node-negative cutaneous melanoma. Ann Surg Oncol 2013;20(2):680–8.

[16] Agrawal A, Civantos FJ, Brumund KT, Chepeha DB, Hall NC, Carroll WR, et al. [(99m)Tc] Tilmanocept accurately detects sentinel lymph nodes and predicts node pathology status in patients with oral squamous cell carcinoma of the head and neck: results of a phase III multiinstitutional trial. Ann Surg Oncol 2015;22(11):3708–15.

[17] Haerle SK, Hany TF, Strobel K, Sidler D, Stoeckli SJ. Is there an additional value of SPECT/CT over planar lymphoscintigraphy for sentinel node mapping in oral/oropharyngeal squamous cell carcinoma? Ann Surg Oncol 2009;16(11):3118–24.

[18] Vermeeren L, van der Ploeg IM, Olmos RA, Meinhardt W, Klop WM, Kroon BB, et al. SPECT/CT for preoperative sentinel node localization. J Surg Oncol 2010;101(2):184–90.

[19] Giammarile F, Schilling C, Gnanasegaran G, Bal C, Oyen WJG, Rubello D, et al. The EANM practical guidelines for sentinel lymph node localisation in oral cavity squamous cell carcinoma. Eur J Nucl Med Mol Imaging 2019;46(3):623–37.

[20] Tsujimoto M, Nakabayashi K, Yoshidome K, Kaneko T, Iwase T, Akiyama F, et al. One-step nucleic acid amplification for intraoperative detection of lymph node metastasis in breast cancer patients. Clin Cancer Res 2007;13(16):4807–16.

[21] Shimazu K, Tanei T, Tamaki Y, Saeki T, Osaki A, Hasebe T, et al. Performance of a new system using a one-step nucleic acid amplification assay for detecting lymph node metastases in breast cancer. Med Oncol (Northwood, London, England) 2019;36(6):54.

[22] Tamaki Y. One-step nucleic acid amplification assay (OSNA) for sentinel lymph node biopsy. Breast Cancer 2015;22(3):230–4.

[23] Wild JB, Iqbal N, Francombe J, Papettas T, Sanders DS, Ramcharan S. Is it time for one-step nucleic acid amplification (OSNA) in colorectal cancer? A systematic review and meta-analysis. Tech Coloproctol 2017;21(9):693–9.

[24] Yamamoto H, Tomita N, Inomata M, Furuhata T, Miyake Y, Noura S, et al. OSNA-assisted molecular staging in colorectal cancer: a prospective multicenter trial in Japan. Ann Surg Oncol 2016;23(2):391–6.

[25] Goda H, Nakashiro KI, Oka R, Tanaka H, Wakisaka H, Hato N, et al. One-step nucleic acid amplification for detecting lymph node metastasis of head and neck squamous cell carcinoma. Oral Oncol 2012;48(10):958–63.

[26] Peigné L, Godey F, Le Gallo M, Le Gall F, Fautrel A, Morcet J, et al. One-step nucleic acid amplification for detecting lymph node metastasis of head and neck squamous cell carcinoma. Oral Oncol 2020;102:104553.

[27] Matsuzuka T, Takahashi K, Kawakita D, Kohno N, Nagafuji H, Yamauchi K, et al. Intraoperative molecular assessment for lymph node metastasis in head and neck squamous cell carcinoma using one-step nucleic acid amplification (OSNA) assay. Ann Surg Oncol 2012;19(12):3865–70.

[28] Suzuki M, Matsuzuka T, Hashimoto Y, Ikeda M, Saijo S, Omori K. Diagnostic potential of 1-step nucleic acid amplification assay in patients with head and neck squamous cell carcinoma based on CK19 expression in a primary lesion. Head Neck 2016;38(Suppl 1):E239–45.

[29] Yamauchi K, Fujioka Y, Kogashiwa Y, Kohno N. Quantitative expression study of four cytokeratins and p63 in squamous cell carcinoma of the tongue: suitability for sentinel node navigation surgery using one-step nucleic acid amplification. J Clin Pathol 2011;64(10):875–9.

162 Inflammation and oral cancer

[30] Shaw R, Christensen A, Java K, Maddani RE, Liloglou T, Asterios T, et al. Intraoperative sentinel lymph node evaluation: implications of cytokeratin 19 expression for the adoption of OSNA in oral squamous cell carcinoma. Ann Surg Oncol 2016;23(12):4042–8.

[31] Noorlag R, van Es RJJ, de Bree R, Willems SM. Cytokeratin 19 expression in early oral squamous cell carcinoma and their metastasis: inadequate biomarker for one-step nucleic acid amplification implementation in sentinel lymph node biopsy procedure. Head Neck 2017;39(9):1864–8.

[32] Shinohara M, Harada T, Nakamura S, Oka M, Tashiro H. Heterotopic salivary gland tissue in lymph nodes of the cervical region. Int J Oral Maxillofac Surg 1992;21(3):166–71.

[33] Rekhtman N, Ang DC, Sima CS, Travis WD, Moreira AL. Immunohistochemical algorithm for differentiation of lung adenocarcinoma and squamous cell carcinoma based on large series of whole-tissue sections with validation in small specimens. Modern Pathol 2011;24(10):1348–59.

[34] Candi E, Rufini A, Terrinoni A, Dinsdale D, Ranalli M, Paradisi A, et al. Differential roles of p63 isoforms in epidermal development: selective genetic complementation in p63 null mice. Cell Death Differ 2006;13(6):1037–47.

[35] Carvalho JC, Thomas DG, McHugh JB, Shah RB, Kunju LP. p63, CK7, PAX8 and INI-1: an optimal immunohistochemical panel to distinguish poorly differentiated urothelial cell carcinoma from high-grade tumours of the renal collecting system. Histopathology 2012;60(4):597–608.

[36] Mukhopadhyay S, Katzenstein AL. Subclassification of non-small cell lung carcinomas lacking morphologic differentiation on biopsy specimens: Utility of an immunohistochemical panel containing TTF-1, napsin A, p63, and CK5/6. Am J Surg Pathol 2011;35(1):15–25.

[37] Ishida K, Tomita H, Kanayama T, Noguchi K, Niwa A, Kawaguchi M, et al. Specific deletion of p16(INK4a) with retention of p19(ARF) enhances the development of invasive oral squamous cell carcinoma. Am J Pathol 2020;190(6):1332–42.

[38] Oka R, Nakashiro K, Goda H, Iwamoto K, Tokuzen N, Hamakawa H. Annexin A8 is a novel molecular marker for detecting lymph node metastasis in oral squamous cell carcinoma. Oncotarget 2016;7(4):4882–9.

[39] Ferris RL, Xi L, Raja S, Hunt JL, Wang J, Gooding WE, et al. Molecular staging of cervical lymph nodes in squamous cell carcinoma of the head and neck. Cancer Res 2005;65(6):2147–56.

[40] Lizarbe MA, Barrasa JI, Olmo N, Gavilanes F, Turnay J. Annexin-phospholipid interactions. Functional implications. Int J Mol Sci 2013;14(2):2652–83.

[41] Solassol J, Burcia V, Costes V, Lacombe J, Mange A, Barbotte E, et al. Pemphigus vulgaris antigen mRNA quantification for the staging of sentinel lymph nodes in head and neck cancer. Br J Cancer 2010;102(1):181–7.

[42] Shores CG, Yin X, Funkhouser W, Yarbrough W. Clinical evaluation of a new molecular method for detection of micrometastases in head and neck squamous cell carcinoma. Arch Otolaryngol Head Neck Surg 2004;130(8):937–42.

[43] Elsheikh MN, Rinaldo A, Hamakawa H, Mahfouz ME, Rodrigo JP, Brennan J, et al. Importance of molecular analysis in detecting cervical lymph node metastasis in head and neck squamous cell carcinoma. Head Neck 2006;28(9):842–9.

[44] Civantos FJ, Zitsch RP, Schuller DE, Agrawal A, Smith RB, Nason R, et al. Sentinel lymph node biopsy accurately stages the regional lymph nodes for T1-T2 oral squamous cell carcinomas: results of a prospective multi-institutional trial. J Clin Oncol 2010;28(8):1395–400.

Standardization of sentinel lymph node biopsy in oral cancer **Chapter | 8** **163**

[45] Alkureishi LW, Ross GL, Shoaib T, Soutar DS, Robertson AG, Thompson R, et al. Sentinel node biopsy in head and neck squamous cell cancer: 5-year follow-up of a European multi-center trial. Ann Surg Oncol 2010;17(9):2459–64.

[46] Schilling C, Stoeckli SJ, Haerle SK, Broglie MA, Huber GF, Sorensen JA, et al. Sentinel European Node Trial (SENT): 3-year results of sentinel node biopsy in oral cancer. Eur J Cancer 2015;51(18):2777–84.

[47] Moya-Plana A, Aupérin A, Guerlain J, Gorphe P, Casiraghi O, Mamelle G, et al. Sentinel node biopsy in early oral squamous cell carcinomas: long-term follow-up and nodal failure analysis. Oral Oncol 2018;82:187–94.

Chapter 9

Overview of radiotherapy for oral cavity cancer

Chiyoko Makita[a], Masaya Ito[a], Hirota Takano[b], Tomoyasu Kumano[a], and Masayuki Matsuo[a]
[a]*Department of Radiology, Gifu University Graduate School of Medicine, Gifu, Japan,*
[b]*Department of Radiotherapy, Gifu University Graduate School of Medicine, Gifu, Japan*

Abbreviations

PORT   postoperative radiotherapy
POCRT postoperative chemoradiotherapy
IMRT   intensity-modulated radiation therapy
IGRT   image-guided radiation therapy
PBT   proton beam therapy
CIT   carbon ion therapy

1. Introduction

Oral cavity cancer involves the lip, floor of the mouth, anterior two-thirds of the tongue, lower alveolar ridge, upper alveolar ridge, retromolar gingiva, hard palate, and buccal mucosa. Squamous cell carcinoma (SCC) is the predominant malignancy, whereas minor salivary gland cancers and sarcomas are less common. The tumor, node, metastasis (TNM) staging system of the Union for International Cancer Control (UICC) and the American Joint Committee on Cancer (AJCC) are used for the classification of oral cavity cancer.

Generally, surgical resection is the standard treatment for patients with resectable tumors. In an effort to improve the cure rates and functional outcomes, radiotherapy has been an essential tool. Definitive radiotherapy alone remains a standard option for patients with stage I to II diseases. Selected high-risk patients with locally advanced disease receive postoperative radiotherapy, with or without concurrent chemotherapy, following primary surgical resection with the aim of improving locoregional control and survival. For patients with unresectable tumors, radiotherapy is adopted as a radical or palliative treatment. In recent years, particle beam therapy has been found to have potential for oral cavity cancer treatment.

The role of radiotherapy in oral cavity cancer treatment will be reviewed here.

Inflammation and Oral Cancer. https://doi.org/10.1016/B978-0-323-88526-3.00009-9
Copyright © 2022 Elsevier Inc. All rights reserved.

165

166 Inflammation and oral cancer

2. General principles

2.1 Definitive radiotherapy for early-stage oral cavity cancer

Transoral or transcervical surgery is generally the first choice for patients with early-stage oral cavity cancer. Definitive radiotherapy has not been compared with primary surgery in randomized trials; however, it is an option for patients who cannot tolerate surgery or for whom surgery would result in particularly severe functional impairment. The general principle of oral cavity cancer treatment is presented in Table 1.

Both external beam radiotherapy (EBRT) and brachytherapy can play a role in the management of early-stage oral cavity cancer. Brachytherapy provides a highly localized dose radiation with a rapid falloff to minimize exposure of normal tissues [1]. This technique alone is not capable of treating regional lymph nodes, and therefore is only appropriate to use in selected early-stage patients with very low risk of occult nodal involvement.

EBRT is adopted as the primary treatment when regional lymph nodes are at a high-risk of metastases, and brachytherapy may be added to enhance treatment for the primary tumor. For patients with stage I oral tongue cancer > 3 mm and stage II oral cavity cancer, elective treatment of the neck is recommended [2, 3].

2.2 Postoperative radiotherapy and chemoradiotherapy

The prognosis for locally advanced SCC of the head and neck is poor. Since Fletcher et al. reported on postoperative radiotherapy (PORT) in 1970 [4], PORT has become the standard treatment for patients with resected locoregionally advanced oral cavity cancer in cases with a significant risk of locoregional recurrence. The risk factors associated with a particularly increased risk of recurrence include extranodal extension (ENE), positive margins, multiple nodal disease, nodal metastases > 3 cm, nodal disease level IV or V, perineural invasion, or vascular invasion [5–11]. The exception would be patients with negative resection margin of a thin primary lesion and a single metastatic lymph node without ENE. In this case, PORT is not performed, as the local recurrence rate is 10%, and the 5-year survival rate is 83%, even with surgery alone [12]. PORT has been performed as postoperative adjuvant therapy for patients with risk factors. However, it has been reported that when there are two or more ECE

TABLE 1 General principle of treatment for oral cavity cancer.

	Stage I	Stage II	Stage III/IV	
			Resectable	Unresectable
Treatment	Surgery or radiotherapy (brachytherapy)		Surgery ± PO(C)RT	Definitive radiotherapy

and recurrence risk factors, the 5-year local recurrence rate is 32%, the distant metastasis rate is 25%, and the 5-year survival rate is 42%, even with PORT alone [13, 14]. The therapeutic effect is considered to be insufficient with PORT alone; thus, treatment development has been promoted.

An integrated analysis of the RTOG85-03 and RTOG88-24 trials was conducted at RTOG, a group centered in the United States [9]. According to these trials, patients with recurrence risk factors, such as microscopic positive margin, ECE, and multiple lymph node metastases, have a lower 5-year local recurrence rate and 5-year survival rate compared with patients without any of these factors. Consequently, these three risk factors are considered to be important for the improvement of prognosis. In EORTC (European Organization for Research and Treatment of Cancer), a group mainly in Europe, in addition to microscopic positive margin and ECE, stage III/IV, perineural infiltration, level IV/V lymph node metastasis, and vascular embolism are considered to be risk factors for recurrence [15]. In 2004, the key trials, EORTC2291 and RTOG95-01, were reported with different definitions of recurrence risk factors between EORTC and RTOG [15, 16]. Both studies demonstrated that postoperative chemoradiotherapy (POCRT) is useful for patients at a high-risk of recurrence. As a result of these two integrated analyses, in 2005, the benefit of POCRT with cisplatin was shown to be even greater in patients with either microscopic positive margins or ECE, which are common factors in both trials.

Since this study, PORT as radiation monotherapy has been performed for patients with multiple lymph node metastases as an intermediate risk factor. There is no large-scale metaanalysis of POCRT, but a metaanalysis that includes the results of four phase III trials demonstrates that chemoradiotherapy significantly contributes to survival compared with radiotherapy alone (relative risk 0.80; 95% CI 0.71–0.90; $P = 0.0002$) [17]. These studies are for SCC of the head and neck, but oral cavity cancer is also included. Based on the above, the standard postoperative adjuvant therapy for oral cavity cancer patients with recurrence risk factors, such as positive margin and ECE, is chemoradiotherapy, and the combination regimen is cisplatin $100 \, \text{mg/m}^2$ [18, 19]. The clinical benefits of a weekly dose of cisplatin at $40 \, \text{mg/m}^2$ were confirmed by an open-label phase II/III trial (JCOG1008) in Japan [20]. In this study, the patients were randomly assigned to POCRT (66 Gy) with either weekly cisplatin ($40 \, \text{mg/m}^2$ weekly for 6–7 doses) or bolus cisplatin ($100 \, \text{mg/m}^2$ every 3 weeks). Weekly cisplatin had a more favorable toxicity profile compared with bolus cisplatin, which is associated with severe acute and late toxicities [21].

The appropriate radiation dose is 60–66 Gy, and the radiation field should basically cover the whole neck [22–25]. In addition, it is recommended that treatment be completed within 56 days after surgery and within 100 days of package time [26, 27]. If the irradiation field is insufficient, the rate of recurrence in the local area significantly increases, and the extension of the treatment period leads to a poor treatment result. The results of PORT, with or without chemotherapy, are presented in Table 2 [15, 28–31]. The IMRT technique and the example of postoperative case will be described in a later section.

TABLE 2 Treatment outcome for patients with HNSCC treated surgery and PORT or POCRT.

Author	Year	Study design	No. of patients	Rate of oral cavity cancer (%)	Treatment	RT (Gy)	Chemotherapy	PFS (year)	OS (year)
Bachaud	1996	Prospective Phase 3	39	NA	CRT	54–74	w-CDDP	45 (5)	36 (5)
			44	NA	RT		None	23 (5)	13 (5)
Al-Sarraf	1997	RTOG88-24	51	15.7	CRT	60	3w-CDDP	43 (3)	48 (3)
Bernier	2004	EORTC 22931 Phase 3	167	25	CRT	66	3w-CDDP	47 (5)	53 (5)
			167	28	RT		None	36 (5)	40 (5)
Cooper	2004	RTOG 95-01 Phase 3	206	24	CRT	60–66	3w-CDDP	47 (3)	56 (3)
			210	30	RT		None	36 (3)	47 (3)
Franchin	2009	Prospective Phase 2	142	31	CRT	66	3w-CDDP	82 (5)	68 (5)

2.3 Primary radiotherapy or chemoradiotherapy without surgery

Functional organ preservation approaches are widely used for patients with locoregionally advanced oropharyngeal, hypopharyngeal, and laryngeal cancers. However, this approach has not been widely applied to patients with oral cavity cancer. Although data are more limited, no survival benefit has been demonstrated in locally advanced oral cavity cancer, and there are concerns over increased toxicity [32–35]. A randomized trial of surgery using PORT or POCRT compared with definitive chemoradiation for locally advanced head and neck cancer found that while survival rates were similar overall between the groups, a subset analysis for oral cavity cancer specifically demonstrated a survival advantage with surgery [36]. Using the National Cancer Database, the effectiveness of organ preservation using concurrent chemoradiotherapy was compared with that of surgery followed by PORT using propensity score matching analysis. The surgery group was associated with improved survival for locally advanced oral cavity SCC, especially in T3–T4a disease [37].

In recent years, it has been reported that the combination of superselective intra-arterial infusion chemotherapy and radiotherapy is effective as a function-preserving therapy [38–41]. Although sufficient evidence has not been demonstrated, this treatment method is expected to preserve organ function as an alternative treatment to surgery.

Traditionally, radiotherapy alone has been performed for unresectable cases, but it has a local recurrence rate of about 70% and the overall survival rate is very poor, at less than 25% [42]. Chemoradiotherapy demonstrated a better control rate of the primary lesion and neck compared with radiotherapy alone [36, 43] and, therefore, chemoradiotherapy has become the standard treatment for unresectable diseases [42, 44]. However, because 38%–89% of patients have grade 3 or higher adverse events, such as mucositis, dermatitis, and myelotoxicity [42, 44–46], the use of platinum-based chemotherapy is recommended. The standard regimen is cisplatin $100\,\text{mg/m}^2$ every 3 weeks and $70\,\text{Gy}$ in 35 fractions. Effective supportive care is essential to complete the treatment [46, 47].

2.4 Palliative radiotherapy

It is widely recognized that palliative radiotherapy provides effective palliation and improved quality of life in patients with advanced incurable oral cavity cancer with distant metastasis, such as brain or bone metastasis [48, 49]. The common symptoms of primary tumor and/or regional lymph nodes include pain, dysphagia, odynophagia, otalgia, hoarseness, and cough. There can be a significant overlap between symptoms and treatment toxicity. The role of radiotherapy in palliation with advanced head and neck cancer is not clear. However, several retrospective case series, case–control study, and single-arm prospective trials confirm that its use is associated with an improvement in outcome [50–54]. Although the optimum dose-fraction schedule is still unclear, the evidence

170 Inflammation and oral cancer

favors a short-course fractionation regimen (e.g., 30 Gy in 10 fractions) or cyclical treatment (QUAD-SHOT) compared with a single-fraction or protracted treatment course [55–57]. Nevertheless, it is recommended to establish proper criteria for palliative radiotherapy in which the primary endpoint is symptom relief and not survival.

2.5 Management for the toxicities of radiotherapy

When chemoradiotherapy is performed for locally advanced head and neck cancer, mucositis occurs in 89%–100% patients (34%–57% of grade 3 and above) in the irradiation field and also causes dysphagia, dysgeusia, and dry mouth [58]. Even if the pain associated with mucositis is actively controlled with acetaminophen or opioids, dysphagia, dysgeusia, and dry mouth often affect oral intake, nutritional status, and weight loss. It is known that these symptoms causes unscheduled pauses in chemoradiotherapy and prolong the treatment period, which adversely influences the therapeutic effect [59]. Thus, nutritional management during chemoradiotherapy in oral cavity cancer is very important. If it is expected that the condition will be prolonged (4 weeks or more), it is recommended to use nutritional support in patients, with a method such as percutaneous endoscopic gastrostomy (PEG), which can be employed for a long time, rather than parenteral nutrition [60].

Late toxicity presents months to years after the completion of treatment and is often permanent. Late complications may include skin and soft tissue atrophy and fibrosis, osteoradionecrosis, lymphedema, and trismus. Xerostomia commonly occurs during the course of radiotherapy and persists after treatment. A mean radiation doses to the parotid glands greater than 24–26 Gy causes permanent damage to the parotid glands, which typically results in more than 75% reduction in salivary gland function [61, 62]. Xerostomia may improve after treatment with modern RT techniques, such as intensity-modulated radiation therapy (IMRT) [63]. However, the improvement of xerostomia in patients with oral cavity cancers may be more limited compared with other head and neck cancers due to high doses of radiation to the submandibular glands and large volumes of oral mucosa and minor salivary glands in the treatment fields.

3. Radiation technique

3.1 EBRT

Brachytherapy is delivered directly to the tumor, close to the source, and the dose distribution demonstrates a rapid dose falloff at target periphery, limiting the dose exposure of the surrounding tissues. In addition, there is no need for additional uncertainty margins such as setup or organ motion uncertainties around the clinical target volume (CTV). Treatment can be delivered within a few days, unlike fractionated EBRT. Brachytherapy, either alone or as a supplement to EBRT, can yield a high tumor control rate with low toxicity profiles in selected patients [1].

Low-dose-rate (LDR) brachytherapy has been employed worldwide owing to its superior outcomes. In Japan, ^{137}Cs, ^{198}Au, or ^{192}Ir are generally used for LDR interstitial radiotherapy. With the advent of technology, high-dose-rate (HDR) brachytherapy has enabled healthcare providers to avoid radiation exposure. Brachytherapy is usually delivered using remote-controlled afterloaders that insert radioactive (^{192}Ir) sources. This therapy has been used for the treatment of many types of cancer, such as gynecological cancer, breast cancer, and prostate cancer.

Brachytherapy alone is recommended for T1N0 and T2N0 tumors <4cm (3) [64, 65]. For tumors >3–4cm or N1 lesions, although surgery is often preferred, brachytherapy can be delivered to the neck and oral cavity as a boost after 40–45 Gy of EBRT [66]. In general, the local control rate is 72%–93% for T1 and T2N0 tumors treated with brachytherapy alone [3, 64, 65, 67]. Approximately 10%–30% of patients may develop soft tissue necrosis within the implant volume. Osteoradionecrosis may occur in 5%–10% of cases [68]. Recently, image-guided brachytherapy has been attempted with computed tomography (CT), magnetic resonance imaging (MRI), and intraoperative ultrasonography to optimize dose distribution [69].

As a very high dose is delivered, acute reactions, such as inflammation, are frequent, and invasive treatment causes a risk of infection and perioperative pain. However, brachytherapy contributes to good long-term functional outcomes with the potential for less fibrosis compared with EBRT.

3.2 Brachytherapy

External beam radiotherapy (EBRT) is the most common type of radiotherpay used for cancer treatment. A medical line accelerator (LINAC) is commonly used for EBRT for patients. It delivers high-energy X-rays or electrons to the region of the patient's tumor. The first step in radiotherapy planning for malignant tumors is the identification of the required volume in a three-dimensional manner. When scanning the patient in the radiation treatment position, CT provides the anatomy, geometry, and electron-density information required for all aspects of planning. CT provides the location of the tumor and normal tissues as well as 3D anatomic information; it also provides the tissue heterogeneity information necessary for accurate therapeutic dose calculations [70]. These volumes are gross tumor volume (GTV), clinical target volume (CTV), planning target volume (PTV), and organs at risk (OAR). GTV and CTV are volumes in which tumor progression is confirmed or suspected, whereas OAR is a normal tissue, yielding the so-called planning organ at risk volume (PRV).

GTV is the macroscopic confirmation of the progression and presence of tumor. It consists of primary lesion (GTV primary) and potentially metastatic lymphadenopathy (GTV nodal). In curative treatment, the entire GTV must be given a sufficient dose. GTV might not be identified even with postoperative or prophylactic radiation. In determining GTV, the requirements for staging tumors based on TNM (UICC) must be met. CTV is a tissue volume that

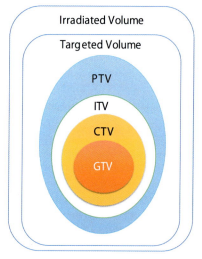

FIG. 1 The ICRU 62 measurement volume. *GTV*, gross tumor volume; *CTV*, clinical target volume; *ITV*, internal target volume; and *PTV*, planning target volume.

contains a demonstrable GTV and subclinical microscopic malignant disease. To perform curative treatment, it is important to administer a sufficient dose to the CTV as well as GTV. PTV is a geometric concept and is defined to select appropriate beam sizes and arrangements, including the setup margin and internal margin, to ensure that the prescribed dose is actually delivered to the CTV. These terminologies were standardized by the International Commission on Radiation Units (ICRU) as ICRU reports 50 and 62 [71]. Fig. 1 presents the ICRU 62 measurement volume. The treatment targets are determined appropriately by a radiation oncologist according to diagnostic images, such as CT, MRI, and positron emission tomography, simulation images, clinical examination, knowledge of cancer biology, and clinical evidences.

Conformal therapy is described as a strategy for conforming the high dose radiation region to the target volume while minimizing the radiation dose to the normal tissues. Three-dimensional conformal radiation therapy (3D CRT) usually implies a CT-based treatment plan. Treatment planning requires consideration of the total dose of radiation required to treat a tumor and balancing this against the potential damage to the normal tissues. The radiation oncologist calculates and optimizes the radiation dose received by the tumor as well as the adjacent normal tissues.

3.3 IMRT

Intensity-modulated radiation therapy (IMRT) is developed from 3D CRT, but changes the strength of some of the beams in certain areas. This enables stronger doses to be delivered to certain parts of the tumor and helps alleviate

Overview of radiotherapy for oral cavity cancer **Chapter | 9 173**

damage to surrounding normal body tissues. IMRT was developed in the late 1980s [72–74]. Even 3D CRT may fail to achieve the correct fluence and dose distributions, but IMRT, using an inverse planning algorithm in which the desired clinical and dosimetric objectives are mathematically stated, may provide excellent dose distributions closely following the shapes and boundaries of the target and relevant OARs in three dimensions.

IMRT has been shown to be useful for reducing long-term morbidity due to head and neck cancer by reducing the doses to critical organs, such as the salivary glands, spinal cord, brain stem, temporal lobes, and auditory and optic structures, without compromising locoregional control. Varying the radiation dose administered within each beam enables IMRT to simultaneously treat multiple areas within the target to different dose levels, thus providing simultaneous integrated boost (SIB) [75]. With SIB-IMRT, all target volumes are irradiated during every radiation session. However, with a properly shaped dose gradient, a higher dose per fraction (e.g., 2 Gy) is delivered to the HR-CTV (high-risk) area, which encompasses the gross tumor (primary lesion and involved nodes), while a lower elective dose (e.g., 1.6 Gy per fraction) is delivered to the subclinical disease target volume, and, when desired, an intermediate dose (e.g., 1.8 Gy per fraction) can be delivered to the intermediate volumes surrounding the HR-CTV location. Daily radiotherapy at a dose of 2 Gy to gross tumor on a 5-day-per-week schedule to a total dose of 66–70 Gy in 7 weeks is the standard definitive treatment for head and neck cancer.

Furthermore, IMRT has succeeded in reducing the frequency of late toxicities, such as parotid gland disorders, while maintaining local control. A meta-analysis of randomized trials comparing IMRT with conventional radiotherapy reported a reduction in grade 2 or higher salivary gland disorders (HR 0.7, 95% CI 0.66–0.87, $P < 0.0001$) [76]. It has also been reported that IMRT is significantly better in improving QOL and dysphagia and that the usefulness of IMRT in the head and neck region is extremely high [77–79]. Fig. 2 presents the comparison of the dose distributions of 3D CRT and IMRT. Fig. 3 presents an example of a postoperative case with ECE. The patient, with pT2N2bM0 SCC of the left lateral tongue, was treated with partial glossectomy with free-flap reconstruction and left selective neck dissection, followed by adjuvant chemoradiotherapy to the primary site and bilateral neck. PTV subclinical (blue line: lower elective dose as subclinical volume), PTV high risk (orange line: intermediate volume), and PTV boost (red line: higher dose as a high-risk volume with ECE) were prescribed 52.8 Gy/33fr (1.6 Gy per fraction), 59.4 Gy/33fr (1.8 Gy per fraction), and 66 Gy/33fr (2 Gy per fraction), respectively.

3.4 IGRT

Image-guided radiation therapy (IGRT) is a form of 3D CRT where imaging scans, such as CT scan, are performed prior to each treatment. This allows the radiation oncologist to adjust the position of the patient or refocus the radiation

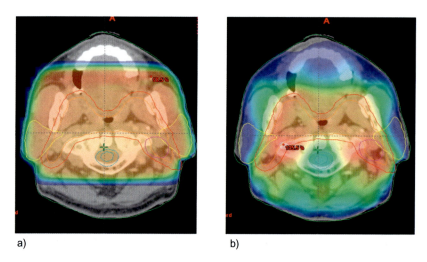

FIG. 2 Dosimetric comparison of three-dimensional conformal radiotherapy (3D CRT) and intensity-modulated radiotherapy (IMRT). (A) 3D CRT and (B) IMRT.

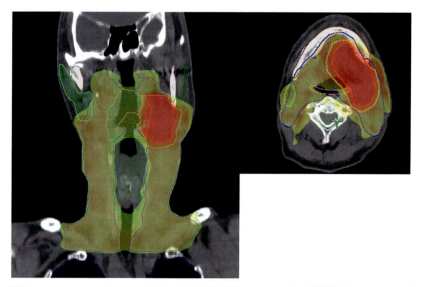

FIG. 3 An example of a postoperative case with ECE: patient with pT2N2bM0 squamous cell carcinoma of the left lateral tongue.

as needed to ensure that the radiation beams are exactly focused on the tumor and that exposure to normal tissues is limited. IGRT is considered to be indispensable for performing high-precision radiotherapy. The addition of IGRT has led to a decrease in the PTV for radiotherapy, which has the potential to decrease normal tissue exposure without compromising locoregional control.

The main aims are to reduce the setup and internal margins and to account for target volume changes during radiotherapy, such as decrease in tumor volume or weight loss. IGRT is not an IMRT technique; however, it enables IMRT to be delivered more accurately [80].

3.5 Particle beam therapy

Particle beam therapy involves the clinical use of ion beams (proton or carbon ions) to treat patients with malignant tumors. It is currently the most advanced form of radiotherapy. The physical advantage of particle beam therapy is the favorable depth-dose distribution compared with conventional X-ray. While photon dose decreases exponentially as a function of the depth, swift heavy ions initially deposit a trivial amount of energy (plateau), but their energy loss per unit track increases with depth, reaching a sharp maximum, which is called the Bragg peak (Fig. 4). In addition, particle beam therapy has several biological advantages compared with conventional photon radiotherapy. Heavy-ion particles, such as carbon ions, have higher linear energy transfer (LET), which inflicts more damage to tissue through direct DNA double-strand breaks [81]. Currently, proton and carbon ions are utilized in clinical settings. Proton has a low LET; however, carbon ion is more expensive than proton beam therapy (PBT) as large synchrotrons are needed to accelerate heavy ions. Therefore, PBT has recently attracted considerable attention worldwide, including Japan.

PBT is also characterized by a rapid fall-off at the distal end of the Bragg peak. The unique physical characteristics of protons make them a promising

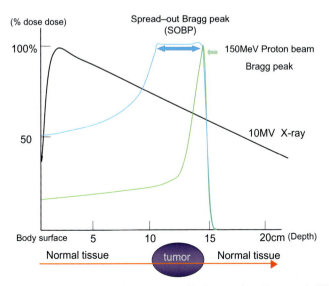

FIG. 4 Comparison of the X-ray depth-dose curve with the spread-out Bragg peak (SOBP) used in proton beam delivery for clinical treatment.

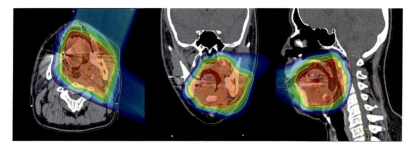

FIG. 5 The case of adenoid cystic carcinoma of the basal tongue treated with proton beam therapy using a broad beam.

option for the treatment of advanced head and neck cancer, with the potential to spare normal tissues from radiation damage and safely escalate the dose of radiation.

In oral cavity cancer, particle beam therapy is recommended for the treatment of non-SCC and locally advanced cancer. Oral non-SCC could be treated effectively, with acceptable toxicity, by carbon ion therapy. Ikawa et al. analyzed 76 patients with oral non-SCC treated across four institutions in Japan from 2004 to 2014. A total of 46 patients had salivary gland carcinoma, and 27 had mucosal melanoma. With a median follow up of 31.1 months, the 3-year LC, PFS, and OS were 86.6%, 63.1%, and 78.4%, respectively. A total of 13 patients had late grade 3 or higher toxicity, with 9 patients having grade 3 osteoradionecrosis. There were no grade 5 toxicities [82]. Charged particle beam re-irradiation also provided superior outcome compared with photon radiotherapy for recurrent head and neck cancer [83]. Excellent treatment results for stage III–IVB SCC of the tongue and locally advanced adenoid cystic carcinoma of the base of the tongue have been reported for proton beam therapy in combination with intra-arterial infusion chemotherapy from Japan [40, 84]. Particle beam therapy for oral cavity cancer is expected to develop in the future. Fig. 5 presents a case of adenoid cystic carcinoma of the basal tongue treated with proton beam therapy.

4. Conclusion

The standard treatment for oral cavity cancer is surgery. Brachytherapy is an option for early-stage cancer. Locally advanced tumors with positive surgical margins and extracapsular extension need a combined treatment, including primary surgery followed by POCRT. Definitive radiotherapy and concurrent chemoradiotherapy are typically reserved for patients who are medically inoperable or who have unresectable diseases. In recent years, high-radiation techniques, such as IMRT and IGRT, have demonstrated reduced morbidity. Particle beam therapies, such as CIT and PBT, have also been attempted.

Acknowledgments

We thank all members of our laboratories.

Conflict of interest statement

No potential conflicts of interest were disclosed.

References

[1] Chargari C, Deutsch E, Blanchard P, Gouy S, Martelli H, Guérin F, et al. Brachytherapy: an overview for clinicians. CA Cancer J Clin 2019;69(5):386–401.

[2] Inoue T, Inoue T, Teshima T, Murayama S, Shimizutani K, Fuchiata H, et al. Phase III trial of high and low dose rate interstitial radiotherapy for early oral tongue cancer. Int J Radiat Oncol Biol Phys 1996;36(5):1201–4.

[3] Shibuya H, Hoshina M, Takeda M, Matsumoto S, Suzuki S, Okada N. Brachytherapy for stage I & II oral tongue cancer: an analysis of past cases focusing on control and complications. Int J Radiat Oncol Biol Phys 1993;26(1):51–8.

[4] Fletcher GH, Evers WT. Radiotherapeutic management of surgical recurrences and postoperative residuals in tumors of the head and neck. Radiology 1970;95(1):185–8.

[5] Laramore GE, Scott CB, Schuller DE, Haselow RE, Ervin TJ, Wheeler R, et al. Is a surgical resection leaving positive margins of benefit to the patient with locally advanced squamous cell carcinoma of the head and neck: a comparative study using the intergroup study 0034 and the radiation therapy oncology group head and neck data. Int J Radiat Oncol Biol Phys 1993;27(5):1011–6.

[6] Jacobs JR, Ahmad K, Casiano R, Schuller DE, Scott C, Laramore GE, et al. Implications of positive surgical margins. Laryngoscope 1993;103(1 Pt 1):64–8.

[7] Rosenthal DI, Mohamed ASR, Garden AS, Morrison WH, El-Naggar AK, Kamal M, et al. Final report of a prospective randomized trial to evaluate the dose-response relationship for postoperative radiation therapy and pathologic risk groups in patients with head and neck cancer. Int J Radiat Oncol Biol Phys [Internet] 2017;98(5):1002–11. Available from: https://doi.org/10.1016/j.ijrobp.2017.02.218.

[8] Hosni A, Huang SH, Chiu K, Xu W, Su J, Bayley A, et al. Predictors of early recurrence prior to planned postoperative radiation therapy for oral cavity squamous cell carcinoma and outcomes following salvage intensified radiation therapy. Int J Radiat Oncol Biol Phys 2019;103(2):363–73.

[9] Cooper JS, Pajak TF, Forastiere A, Jacobs J, Fu KK, Ang KK, et al. Precisely defining high-risk operable head and neck tumors based on RTOG #85-03 and #88-24: targets for postoperative radiochemotherapy? Head Neck 1998;20(7):588–94.

[10] Langendijk JA, Slotman BJ, Van Der Waal I, Doornaert P, Berkof J, Leemans CR. Risk-group definition by recursive partitioning analysis of patients with squamous cell head and neck carcinoma treated with surgery and postoperative radiotherapy. Cancer 2005;104(7):1408–17.

[11] Soo KC, Carter RL, O'Brien CJ, Barr L, Bliss JM, Shaw HJ. Prognostic implications of perineural spread in squamous carcinomas of the head and neck. Laryngoscope 1986;96(10):1145–8.

[12] Ang KK, Trotti A, Brown BW, Garden AS, Foote RL, Morrison WH, et al. Randomized trial addressing risk features and time factors of surgery plus radiotherapy in advanced head-and-neck cancer. Int J Radiat Oncol Biol Phys 2001;51(3):571–8.

178 Inflammation and oral cancer

[13] Ang KK, Trotti A, Brown BW, Garden AS, Foote RL, Morrison WH, et al. Randomized trial addressing risk features and time head-and-neck cancer. Radiat Oncol 2001;51(3):571–8.

[14] Laramore GE, Scott CB, Al-Sarraf M, Haselow RE, Ervin TJ, Wheeler R, Jacobs JR, et al. Adjuvant chemotherapy for resectable squamous cell carcinomas of the head and neck: report on Intergroup Study 0034. Int J Radiat Oncol Biol Phys 1992;23(4):705–13.

[15] Cooper JS, Pajak TF, Forastiere AA, Jacobs J, Campbell BH, Saxman SB, et al. Postoperative concurrent radiotherapy and chemotherapy for high-risk squamous-cell carcinoma of the head and neck. N Engl J Med 2004;350(19):1937–44.

[16] Bernier J, Domenge C, Bernier J, Domenge C, Ozsahin M, Matuszewska K, Lefèbvre J-L, Greiner RH, Giralt J, Maingon P, Rolland F, Bolla M, et al. Postoperative irradiation with or without concomitant chemotherapy for locally advanced head and neck cancer. N Engl J Med 2004;350(19):1945–52. https://doi.org/10.1056/nejmoa032641.

[17] Bernier J, Cooper JS, Pajak TF, Van Glabbeke M, Bourhis J, Forastiere A, et al. Defining risk levels in locally advanced head and neck cancers: a comparative analysis of concurrent postoperative radiation plus chemotherapy trials of the EORTC (#22931) and RTOG (#9501). Head Neck 2005;27(10):843–50.

[18] Winquist E, Oliver T, Gilbert R. Postoperative chemoradiotherapy for advanced squamous cell carcinoma of the head and neck: a systematic review with meta-analysis. Head Neck 2007;29(1):38–46.

[19] Huang SH, O'Sullivan B. Oral cancer: current role of radiotherapy and chemotherapy. Med Oral Patol Oral Cir Bucal 2013;18(2).

[20] Kunieda F, Kiyota N, Tahara M, Kodaira T, Hayashi R, Ishikura S, et al. Randomized phase II/III trial of post-operative chemoradiotherapy comparing 3-weekly cisplatin with weekly cisplatin in high-risk patients with squamous cell carcinoma of head and neck: Japan clinical oncology group study (JCOG1008). Jpn J Clin Oncol 2014;44(8):770–4.

[21] Szturz P, Wouters K, Kiyota N, Makoto T, Prabhash K. Weekly low-dose versus three-weekly high-dose cisplatin for concurrent chemoradiation in locoregionally advanced non-nasopharyngeal head and neck cancer: a systematic review and meta-analysis of aggregate data. Oncologist 2017;22(9):1022–3.

[22] Makita C, Kodaira T, Daimon T, Tachibana H, Tomita N, Koide Y, et al. Comparisons of the clinical outcomes of different postoperative radiation strategies for treatment of head and neck squamous cell carcinoma. Jpn J Clin Oncol 2017;1–10.

[23] Mohamed ASR, Wong AJ, Fuller CD, Kamal M, Gunn GB, Phan J, et al. Patterns of locoregional failure following post-operative intensity-modulated radiotherapy to oral cavity cancer: quantitative spatial and dosimetric analysis using a deformable image registration workflow. Radiat Oncol 2017;12(1):1–12.

[24] Damast S, Wolden S, Lee N. Marginal recurrences after selective targeting with intensity-modulated radiotherapy for oral tongue cancer. Head Neck 2012;34(6):900–6.

[25] Evans M, Beasley M. Target delineation for postoperative treatment of head and neck cancer. Oral Oncol 2018;86:288–95.

[26] Rosenthal DI, Liu L, Lee JH, Vapiwala N, Chalian AA, Weinstein GS, et al. Importance of the treatment package time in surgery and postoperative radiation therapy for squamous carcinoma of the head and neck. Head Neck 2002;24(2):115–26.

[27] Huang J, Barbera L, Brouwers M, Browman G, Mackillop WJ. Does delay in starting treatment affect the outcomes of radiotherapy? A systematic review. *J Clin Oncol Off J Am Soc Clin Oncol* 2003;21(3):555–63.

[28] Bachaud JM, Cohen-Jonathan E, Alzieu C, David JM, Serrano E, Daly-Schveitzer N. Combined postoperative radiotherapy and weekly cisplatin infusion for locally advanced head and

Overview of radiotherapy for oral cavity cancer Chapter | 9 **179**

neck carcinoma: final report of a randomized trial. Int J Radiat Oncol Biol Phys [Internet] 1996;36(5):999–1004. Available from http://www.ncbi.nlm.nih.gov/pubmed/8985019.

[29] Al-Sarraf M, Pajak TF, Byhardt RW, Beitler JJ, Salter MM, Cooper JS. Postoperative radiotherapy with concurrent cisplatin appears to improve locoregional control of advanced, resectable head and neck cancers: RTOG 88-24. Int J Radiat Oncol Biol Phys 1997;37(4):777–82.

[30] Bernier J, Domenge C, Ozsahin M, Matuszewska K, Lefebvre JL, Greiner RH, et al. Postoperative irradiation with or without concomitant chemotherapy for locally advanced head and neck cancer. New Engl Med 2004;350(19):1945–52.

[31] Franchin G, Minatel E, Politi D, Gobitti C, Talamini R, Vaccher E, et al. Postoperative reduced dose of cisplatin concomitant with radiation therapy in high-risk head and neck squamous cell carcinoma. Cancer 2009;115(11):2464–71.

[32] Turner SL, Slevin NJ, Gupta NK, Swindell R. Radical external beam radiotherapy for 333 squamous carcinomas of the oral cavity—evaluation of late morbidity and a watch policy for the clinically negative neck. Radiother Oncol 1996;41(1):21–9.

[33] Uppaluri R, Campbell KM, Egloff AM, Zolkind P, Skidmore ZL, Nussenbaum B, et al. Neoadjuvant and adjuvant Pembrolizumab in resectable locally advanced, human papillomavirus-unrelated head and neck Cancer: a multicenter, phase II trial. Clin Cancer Res 2020;26(19):5140–52.

[34] Zhong L, Zhang C, Ren G, Guo W, William WNJ, Sun J, et al. Randomized phase III trial of induction chemotherapy with docetaxel, cisplatin, and fluorouracil followed by surgery versus up-front surgery in locally advanced resectable oral squamous cell carcinoma. J Clin Oncol 2013;31(6):744–51.

[35] Licitra L, Grandi C, Guzzo M, Mariani L, Lo Vullo S, Valvo F, et al. Primary chemotherapy in resectable oral cavity squamous cell cancer: a randomized controlled trial. J Clin Oncol Off J Am Soc Clin Oncol 2003;21(2):327–33.

[36] Iyer NG, Tan DSW, Tan VKM, Wang W, Hwang J, Tan NC, et al. Randomized trial comparing surgery and adjuvant radiotherapy versus concurrent chemoradiotherapy in patients with advanced, nonmetastatic squamous cell carcinoma of the head and neck: 10-year update and subset analysis. Cancer 2015;121(10):1599–607.

[37] Spiotto MT, Jefferson G, Wenig B, Markiewicz M, Weichselbaum RR, Koshy M. Differences in survival with surgery and postoperative radiotherapy compared with definitive chemoradiotherapy for oral cavity cancer: a national cancer database analysis. JAMA Otolaryngol Head Neck Surg 2017;143(7):691–9.

[38] Mitsudo K, Shigetomi T, Fujimoto Y, Nishiguchi H, Yamamoto N, Furue H, et al. Organ preservation with daily concurrent chemoradiotherapy using superselective intra-arterial infusion via a superficial temporal artery for T3 and T4 head and neck cancer. Int J Radiat Oncol Biol Phys 2011;79(5):1428–35.

[39] Fuwa N, Kodaira T, Furutani K, Tachibana H, Nakamura T. Treatment results of continuous intra-arterial CBDCA infusion chemotherapy in combination with radiation therapy for locally advanced tongue cancer. Oral Surg Oral Med Oral Pathol Oral Radiol Endod 2008;105(6):714–9.

[40] Takayama K, Kato T, Nakamura T, Azami Y, Ono T, Suzuki M, et al. Proton beam therapy combined with intra-arterial infusion chemotherapy for stage IV adenoid cystic carcinoma of the base of the tongue. Cancers (Basel) 2019;11(10).

[41] Fuwa N, Kodaira T, Furutani K, Tachibana H, Nakamura T, Daimon T. Chemoradiation therapy using radiotherapy, systemic chemotherapy with 5-fluorouracil and nedaplatin, and intra-arterial infusion using carboplatin for locally advanced head and neck cancer—phase II study. Oral Oncol 2007;43(10):1014–20.

180 Inflammation and oral cancer

[42] Adelstein DJ, Li Y, Adams GL, Wagner H, Kish JA, Ensley JF, et al. An intergroup phase III comparison of standard radiation therapy and two schedules of concurrent chemoradiotherapy in patients with unresectable squamous cell head and neck cancer. J Clin Oncol 2003;21(1):92–8.

[43] Robertson AG, Soutar DS, Paul J, Webster M, Leonard AG, Moore KP, et al. Early closure of a randomized trial: surgery and postoperative radiotherapy versus radiotherapy in the management of intra-oral tumours. Clin Oncol (R Coll Radiol) 1998;10(3):155–60.

[44] Argiris A, Karamouzis MV, Raben D, Ferris RL. Head and neck cancer. Lancet (London, England) 2008;371(9625):1695–709.

[45] Wendt TG, Grabenbauer GG, Rödel CM, Thiel HJ, Aydin H, Rohloff R, et al. Simultaneous radiochemotherapy versus radiotherapy alone in advanced head and neck cancer: a randomized multicenter study. J Clin Oncol Off J Am Soc Clin Oncol 1998;16(4):1318–24.

[46] Elting LS, Cooksley C, Chambers M, Cantor SB, Manzullo E, Rubenstein EB. The burdens of cancer therapy: clinical and economic outcomes of chemotherapy-induced mucositis. Cancer 2003;98(7):1531–9.

[47] Kojima Y, Yanamoto S, Umeda M, Kawashita Y, Saito I, Hasegawa T, et al. Relationship between dental status and development of osteoradionecrosis of the jaw: a multicenter retrospective study. Oral Surg Oral Med Oral Pathol Oral Radiol 2017;124(2):139–45.

[48] Khuntia D, Brown P, Li J, Mehta MP. Whole-brain radiotherapy in the management of brain metastasis. J Clin Oncol 2006;24(8):1295–304.

[49] Chow E, Wu JSY, Hoskin P, Coia LR, Bentzen SM, Blitzer PH. International consensus on palliative radiotherapy endpoints for future clinical trials in bone metastases. Radiother Oncol 2002;64(3):275–80.

[50] Lusinchi A, Bourhis J, Wibault P, Le Ridant AM, Eschwege F. Radiation therapy for head and neck cancers in the elderly. Int J Radiat Oncol Biol Phys 1990 Apr;18(4):819–23.

[51] Erkal HS, Mendenhall WM, Amdur RJ, Villaret DB, Stringer SP. Squamous cell carcinomas metastatic to cervical lymph nodes from an unknown head and neck mucosal site treated with radiation therapy with palliative intent. Radiother Oncol J Eur Soc Ther Radiol Oncol 2001;59(3):319–21.

[52] Wendt TG, Wustrow TP, Hartenstein RC, Rohloff R, Trott KR. Accelerated split-course radiotherapy and simultaneous cis-dichlorodiammine-platinum and 5-fluorouracil chemotherapy with folinic acid enhancement for unresectable carcinoma of the head and neck. Radiother Oncol J Eur Soc Ther Radiol Oncol 1987;10(4):277–84.

[53] Minatel E, Gigante M, Franchin G, Gobitti C, Mascarin M, Bujor L, et al. Combined radiotherapy and bleomycin in patients with inoperable head and neck cancer with unfavourable prognostic factors and severe symptoms. Oral Oncol 1998;34(2):119–22.

[54] Paris KJ, Spanos WJJ, Lindberg RD, Jose B, Albrink F. Phase I-II study of multiple daily fractions for palliation of advanced head and neck malignancies. Int J Radiat Oncol Biol Phys 1993;25(4):657–60.

[55] Toya R, Saito T, Yamaguchi K, Matsuyama T, Watakabe T, Matsumoto T, et al. Hypofractionated palliative volumetric modulated arc radiotherapy with the radiation oncology study group 8502 "qUAD shot" regimen for incurable head and neck cancer. Radiat Oncol 2020;15(1):1–8.

[56] Chen AM, Vaughan A, Narayan S, Vijayakumar S. Palliative radiation therapy for head and neck cancer: toward an optimal fractionation scheme. Head Neck 2008;30(12):1586–91.

[57] Lok BH, Jiang G, Gutiontov S, Lanning RM, Sridhara S, Sherman EJ, et al. Palliative head and neck radiotherapy with the RTOG 8502 regimen for incurable primary or metastatic cancers. Oral Oncol 2015;51(10):957–62.

[58] Trotti A, Bellm LA, Epstein JB, Frame D, Fuchs HJ, Gwede CK, et al. Mucositis incidence, severity and associated outcomes in patients with head and neck cancer receiving radiotherapy with or without chemotherapy: a systematic literature review. Radiother Oncol 2003;66(3):253–62.

[59] Fowler JF, Lindstrom MJ. Loss of local control with prolongation in radiotherapy. Int J Radiat Oncol Biol Phys 1992;23(2):457–67.

[60] Martindale RG, McClave SA, Vanek VW, McCarthy M, Roberts P, Taylor B, et al. Guidelines for the provision and assessment of nutrition support therapy in the adult critically ill patient: society of critical care medicine and American society for parenteral and enteral nutrition: executive summary. Crit Care Med 2009;37(5):1757–61.

[61] Braam PM, Roesink JM, Moerland MA, Raaijmakers CPJ, Schipper M, Terhaard CHJ. Long-term parotid gland function after radiotherapy. Int J Radiat Oncol Biol Phys 2005;62(3):659–64.

[62] Deasy JO, Moiseenko V, Marks L, Chao KSC, Nam J, Eisbruch A. Radiotherapy dose-volume effects on salivary gland function. Int J Radiat Oncol Biol Phys 2010;76(3 SUPPL):58–63.

[63] Zhang B, Mo Z, Du W, Wang Y, Liu L, Wei Y. Intensity-modulated radiation therapy versus 2D-RT or 3D-CRT for the treatment of nasopharyngeal carcinoma: a systematic review and meta-analysis. Oral Oncol 2015;51(11):1041–6.

[64] Ichimiya Y, Fuwa N, Kamata M, Kodaira T, Furutani K, Tachibana H, et al. Treatment results of stage I oral tongue cancer with definitive radiotherapy. Oral Oncol 2005;41(5):520–5.

[65] Fujita M, Hirokawa Y, Kashiwado K, Akagi Y, Kashimoto K, Kiriu H, et al. Interstitial brachytherapy for stage I and II squamous cell carcinoma of the oral tongue: factors influencing local control and soft tissue complications. Int J Radiat Oncol Biol Phys 1999;44(4):767–75.

[66] Ihara N, Shibuya H, Yoshimura R, Oota S, Miura M, Watanabe H. Interstitial brachytherapy and neck dissection for stage III squamous cell carcinoma of the mobile tongue. Acta Oncol (Madr) 2005;44(7):709–16.

[67] Akiyama H, Yoshida K, Shimizutani K, Yamazaki H, Koizumi M, Yoshioka Y, et al. Dose reduction trial from 60 Gy in 10 fractions to 54 Gy in 9 fractions schedule in high-dose-rate interstitial brachytherapy for early oral tongue cancer. J Radiat Res 2012;53(5):722–6.

[68] Mazeron J-J, Ardiet J-M, Haie-Méder C, Kovács G, Levendag P, Peiffert D, et al. GEC-ESTRO recommendations for brachytherapy for head and neck squamous cell carcinomas. Radiother Oncol 2009;91(2):150–6.

[69] Nag S. High dose rate brachytherapy: its clinical applications and treatment guidelines. Technol Cancer Res Treat 2004;3(3):269–87.

[70] Aird EGA, Conway J. CT simulation for radiotherapy treatment planning. Br J Radiol 2002;75(900):937–49.

[71] Burnet NG, Noble DJ, Paul A, Whitfield GA, Delorme S. Target volume concepts in radiotherapy and their implications for imaging. Radiologe 2018;58(8):708–21.

[72] Brahme A. Design principles and clinical possibilities with a new generation of radiation therapy equipment: a review. Acta Oncol (Madr) 1987;26(6):403–12.

[73] Brahme A. Optimization of stationary and moving beam radiation therapy techniques. Radiother Oncol J Eur Soc Ther Radiol Oncol 1988;12(2):129–40.

[74] Bortfeld T, Bürkelbach J, Boesecke R, Schlegel W. Methods of image reconstruction from projections applied to conformation radiotherapy. Phys Med Biol 1990;35(10):1423–34.

[75] Mohan R, Wu Q, Manning M, Schmidt-Ullrich R. Radiobiological considerations in the design of fractionation strategies for intensity-modulated radiation therapy of head and neck cancers. Int J Radiat Oncol Biol Phys 2000;46(3):619–30.

182 Inflammation and oral cancer

[76] Marta GN, Silva V, de Andrade CH, de Arruda FF, Hanna SA, Gadia R, et al. Intensity-modulated radiation therapy for head and neck cancer: systematic review and meta-analysis. Radiother Oncol 2014;110(1):9–15.

[77] Tribius S, Raguse M, Voigt C, Münscher A, Gröbe A, Petersen C, et al. Residual deficits in quality of life one year after intensity-modulated radiotherapy for patients with locally advanced head and neck cancer: results of a prospective study. Strahlenther Onkol 2015;191(6):501–10.

[78] Klein J, Livergant J, Ringash J. Health related quality of life in head and neck cancer treated with radiation therapy with or without chemotherapy: a systematic review. Oral Oncol 2014;50(4):254–62.

[79] Staffurth J. A review of the clinical evidence for intensity-modulated radiotherapy. Clin Oncol 2010;22(8):643–57.

[80] Bhide SA, Nutting CM. Advances in radiotherapy for head and neck cancer. Oral Oncol 2010;46(6):439–41.

[81] Huang YW, Pan CY, Hsiao YY, Chao TC, Lee CC, Tung CJ. Monte Carlo simulations of the relative biological effectiveness for DNA double strand breaks from 300 MeV u(-1) carbon-ion beams. Phys Med Biol 2015;60(15):5995–6012.

[82] Ikawa H, Koto M, Demizu Y, Ichi SJ, Suefuji H, Okimoto T, et al. Multicenter study of carbon-ion radiation therapy for nonsquamous cell carcinomas of the oral cavity. Cancer Med 2019;8(12):5482–91.

[83] Yamazaki H, Demizu Y, Okimoto T, Ogita M, Himei K, Nakamura S, et al. Reirradiation for recurrent head and neck cancers using charged particle or photon radiotherapy. Strahlenther Onkol 2017;193(7):525–33.

[84] Takayama K, Nakamura T, Takada A, Makita C, Suzuki M, Azami Y, et al. Treatment results of alternating chemoradiotherapy followed by proton beam therapy boost combined with intra-arterial infusion chemotherapy for stage III–IVB tongue cancer. J Cancer Res Clin Oncol 2016;142(3).

Chapter 10

Management of cancer treatment-induced oral mucositis

Akio Suzuki
Department of Pharmacy, Gifu University Hospital, Gifu, Japan

1. Introduction

Oral mucositis (OM) is an acute inflammation of the oral cavity and is a common and distressing complication that is secondary to cancer treatment, including chemotherapy and/or radiotherapy. The clinical presentation of OM ranges from a general erythematous stomatitis to erosive lesions and overt ulceration. Severe OM is accompanied by pain, odynophagia, dysgeusia and subsequent malnutrition and dehydration, which severely impair patients' quality of life (QOL) [1, 2]. Moreover, severe OM is significantly associated with the development of infection [3]. The presence of severe OM after completion of radiotherapy or chemotherapy is one of the major causes of delayed discharge from hospital [3, 4]. Therefore, the incidence of severe OM increases the cost of care associated with hospitalization and medical management, nutritional support, and management of secondary infection [3–5]. Moreover, severe OM is associated with discontinuation of radiotherapy or a reduction in the chemotherapy dose, which negatively impacts the therapeutic efficacy of treatment [3, 6]. Thus, OM is a dose-limiting toxicity of radiotherapy or chemotherapy in patients receiving cancer treatment.

In the following sections, we review the frequency, mechanism of pathogenesis, and preventive and treatment measures, including oral care, cryotherapy, and a number of agents (vitamin E, allopurinol, glutamine, zinc sucralfate, benzydamine, and palifermin), for cancer treatment-induced OM, including chemotherapy and/or radiotherapy. Additionally, we detail findings on the preventive effects of polaprezinc, a zinc-L-carnosine, for chemotherapy and/or radiotherapy-induced OM, and the impact of this compound on OM.

Inflammation and Oral Cancer. https://doi.org/10.1016/B978-0-323-88526-3.00010-5
Copyright © 2022 Elsevier Inc. All rights reserved.

2. Frequency of oral mucositis

Conventional cytotoxic drugs such as fluorinated pyrimidine, methotrexate, doxorubicin, and actinomycin D are frequently associated with the occurrence of OM, which is observed in about 20%–40% of patients receiving these chemotherapy drugs [7]. The incidence of OM is even more frequently associated with chemotherapy followed by hematopoietic stem cell transplantation (HSCT) and head and neck radiotherapy or radio-chemotherapy, which occur in approximately 80% and 100% of patients, respectively [8, 9]. OM incidence is also as high as 40%–60% following treatment with anticancer targeted therapies such as everolimus and temsirolimus; inhibitors of the mammalian target of rapamycin (mTOR); cetuximab, an epidermal growth factor receptor (EGFR) monoclonal antibody; and afatinib, an EGFR tyrosine kinase inhibitor [10].

3. Pathogenesis of anticancer agent and radiotherapy-induced oral mucositis

Although the precise mechanisms of the pathogenesis of chemotherapy- and/ or radiotherapy-induced OM remain to be clarified, evidence suggests similar mechanisms are responsible for the development of both chemotherapy-induced and radiotherapy-induced OM [11, 12].

A previous study demonstrated elevated levels of nuclear factor κB (NFκB) and proinflammatory cytokines, including tumor necrosis factor-α (TNF-α), interleukin-6 (IL-6), and interleukin-1β (IL-1β), and reduced IL-13 levels in OM following administration of irinotecan, methotrexate, or 5-fluorouracil (5-FU) in a Dark Agouti rat model [13]. Moreover, hamster models in which the cheeks were treated with radiation showed elevated nitric oxide (NO), nitric oxide synthase (iNOS), TNF-α, and IL-1β, and reduced IL-10 levels [14].

In humans, analysis of serum levels of IL-6 in 52 patients undergoing HSCT by Min et al. showed that IL-6 levels at 1 week after HSCT were correlated with severe OM [15]. Silva et al. examined the effects of low-level laser therapy on inflammatory mediator release in patients with chemotherapy-induced OM undergoing HSCT, and found that IL-6, IL-1β, TNF-α, and matrix metalloproteinase-2 (MMP-2) levels in saliva were significantly elevated on day 7 after HSCT [16]. Haddad et al. examined 58 head and neck patients receiving chemoradiotherapy and found elevated serum levels of TNF-α, IL-1β, and IL-6 in patients with OM compared to patients without OM [17].

Functional disruption to the submucosa following chemotherapy and/ or radiotherapy via the actions of reactive oxygen species (ROS) and proinflammatory cytokines and chemotherapy-induced neutropenia facilitates the invasion of microorganisms into the tissue, which can lead to secondary injury to the oral mucosa. While bacterial infections are often associated with secondary disorders of OM, fungi and herpes simplex virus are also causal factors [18, 19].

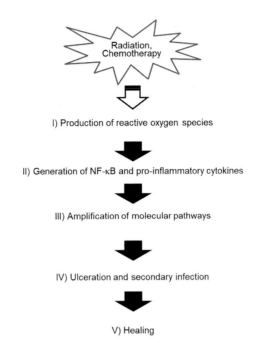

FIG. 1 The developmental process of chemotherapy- and/or radiotherapy-induced oral mucositis.

The developmental process of OM can be divided into five phases: (1) initiation of mucosal injury due to the production of ROS following systemic chemotherapy or radiotherapy; (2) tissue injury and cell death induced by upregulation of NFκB and proinflammatory cytokines such as TNF-α, IL-1, and IL-6; (3) amplification of molecular pathways that lead to mucosal injury by NFκB and proinflammatory cytokines; (4) formation of mucosal ulceration caused by infiltration of inflammatory cells and worsening of symptoms due to secondary infection; and (5) healing mediated by epithelial proliferation and cellular and tissue differentiation (Fig. 1) [11, 20].

4. Management of oral mucositis

4.1 Oral care

Basic oral care, including mechanical cleaning (tooth brushing, flossing), mouth washing (bland rinses), and hydration and lubrication (applying moisturizing agents), is important for maintaining good oral health, reducing the risk of infection and inflammation in the oral cavity, and preventing systemic infection from the oral cavity. The main purpose of basic oral care in the management of OM is to reduce the bacterial load in the oral cavity and prevent secondary disorders of the mucosa due to oral soft tissue infection.

186 Inflammation and oral cancer

The Multinational Association of Supportive Care in Cancer/International Society for Oral Oncology (MASCC/ISOO) guidelines indicate the benefits of implementing multiagent combination oral care protocols for preventing OM during chemotherapy, head and neck radiotherapy, and HSCT [21]. However, the guidelines also suggest that chlorhexidine should not be used to prevent OM in patients receiving head and neck radiotherapy.

4.2 Cryotherapy

Cryotherapy typically involves applying ice or cold water to the oral cavity. Local oral cooling by cryotherapy is thought to cause vasoconstriction in the oral cavity, thereby suppressing the delivery of anticancer agents to oral tissue [22]. A major consideration is to use spherical-shaped ice to prevent oral cavity damage by ice pieces. Alternatively, an intraoral cooling device that generates a cool environment in the oral cavity has been reported to significantly improve toleration compared to ice chips [23]. However, it should be noted that application of oral cryotherapy in oral cancer may reduce the antitumor effects of anticancer agents [22].

The MASCC/ISOO clinical practice guidelines recommend oral cryotherapy for the prevention of OM in patients undergoing autologous HSCT with conditioning regimen protocols and those receiving bolus 5-FU for solid tumors when used during the 30-min chemotherapy infusion [24].

4.3 Pharmacotherapy for protection or treatment of chemotherapy- and/or radiotherapy-induced oral mucositis

4.3.1 Vitamin E

Vitamin E, which refers to α-, β-, γ-, δ-tocopherol and α-, β-, γ-, δ-tocotrienol, has antioxidant and membrane-stabilizing actions, and is thought to interfere with the inflammatory damage caused by reactive oxygen free radicals created in the course of chemotherapy and/or radiotherapy [25–27].

Four randomized controlled trials (RCTs) have reported the effects of vitamin E on chemotherapy and/or radiotherapy-induced OM for various types of cancer treatments [28–31]. Of these, two RCTs reported the treatment effects of vitamin E on OM [28, 29]. Wadleigh et al. showed that among patients receiving chemotherapy for various types of malignancy, the rate of complete resolution of oral lesions was significantly higher among those who received topical vitamin E than those who received placebo (67% for vitamin E group vs 11% for placebo group, $P < 0.025$) [28]. However, the sample size of this study was small ($n = 18$). Similarly, Khurana et al. reported that among 72 children with hematologic malignancies receiving chemotherapy, swish and swallow vitamin E treatment significant improved OM compared to glycerin (control) [29].

The other two RCTs reported the protective effects of vitamin E on OM [30, 31]. A double-blind randomized trial conducted by Ferreira et al. in 54 patients receiving head and neck radiotherapy revealed that vitamin E (400 mg

Management of cancer treatment-induced oral mucositis Chapter | 10 **187**

vitamin E rinse of the oral cavity for 5 min, then swallow) significantly reduced the incidence of Grade 2–3 OM compared to placebo (11% for vitamin E group vs 54% placebo group, $P < 0.001$) [30]. There were no significant differences in survival rate between the two groups. In contrast, Sung et al. reported that there was no significant difference in OM severity with the use of topical vitamin E (800 mg) compared to placebo in 16 children undergoing doxorubicin-containing chemotherapy [31]. However, the sample size of this study was small.

Interestingly, in a randomized study (no control arm), EI-Housseiny et al. reported that topically applied vitamin E significantly improved group grades of OM among 80 children with chemotherapy-induced OM, while systemic administration of vitamin E did not [32].

4.3.2 Allopurinol

Allopurinol, a xanthine oxidase inhibitor, reportedly reduces the production of active oxygen in the oral cavity via inhibition of xanthine oxidase and protease [33]. In the treatment of OM, allopurinol is administered as an ice ball (900 mg/day) or mouthwash (0.1%).

The protective and treatment effects of allopurinol on OM have been primarily evaluated in patients receiving 5-FU-based chemotherapy. However, its effectiveness is controversial. Porta et al. conducted a randomized double-blind, placebo-controlled trial to evaluate the treatment effects of allopurinol mouthwashes on OM in 44 advanced cancer patients receiving 5-FU-containing chemotherapy [34]. Allopurinol mouthwashes significantly resolved OM ($P < 0.001$) and shortened the duration of OM symptoms compared to placebo (4.0 days vs 7.5 days, $P < 0.0005$). Yokomizo et al. investigated the prophylactic efficacy of allopurinol ice balls in patients receiving leucovorin/5-FU therapy. Fifteen of 32 (47%) patients who did not use allopurinol ice balls developed OM, vs only 3 of 20 (10%) patients who received chemotherapy in conjunction with allopurinol ice balls ($P = 0.0187$) [35]. Elzawawy et al. evaluated the prophylactic effects of allopurinol mouthwash in 18 patients with gastrointestinal cancer who had experienced 5-FU-induced OM in previous cycles. They found that allopurinol mouthwash reduced the severity of OM in 15 patients (83.3%) [36].

In contrast, in an RCT of 33 patients with malignant disorders who received 5-FU-based chemotherapy, Panahi et al. demonstrated that allopurinol mouthwash produced no significant differences in the occurrence ($P = 0.256$) or severity ($P = 0.386$) of OM compared to placebo [37]. Moreover, in a phase II trial of cisplatin and 5-FU with allopurinol for recurrent or metastatic carcinoma of the uterine cervix, Weiss et al. reported that allopurinol produced no improvement in treatment-related toxicities, including OM [38]. Accordingly, the protective and treatment effects of allopurinol on cancer treatment-induced OM remain unclear.

4.3.3 Sucralfate

Sucralfate, a nonabsorbable basic aluminum salt of sulfated disaccharide, is often used to treat peptic ulcers. Sucralfate forms an adhesive substance that acts

188 Inflammation and oral cancer

as a protective barrier, and promotes the healing of damaged mucosa by enhancing local prostaglandin production [39].

Several RCTs have examined the preventive effects of sucralfate on OM induced by head and neck radiotherapy, 5-FU-based chemotherapy, and chemotherapy ± total body irradiation (TBI) followed by HSCT [40–42]. One RCT that examined the effectiveness of sucralfate for preventing OM in 44 patients receiving head and neck radiotherapy revealed that sucralfate treatment significantly reduced OM scores compared to placebo (2 ± 1.2 vs 4 ± 0.8, $P = 0.0002$) [40]. Castagna et al. reported the protective effects of sucralfate on OM in 102 patients receiving high-dose chemotherapy ± TBI followed by HSCT. They found that sucralfate tended to reduce the incidence of severe OM compared to placebo, although the effect was not statistically significant (29% vs 47%, $P = 0.07$) [41]. Ala et al. showed that patients with gastrointestinal cancers receiving 5-FU-based chemotherapy and using a sucralfate mouthwash experienced lower frequency and severity of OM (76% vs 38.5%, $P = 0.005$ and 84 vs 38.5%, $P < 0.001$, respectively) and less intense pain (2.5 ± 2.2 vs 5.08 ± 3.82, $P = 0.004$ and 1.33 ± 0.86 vs 4.12 ± 3.5, $P = 0.001$, respectively) compared to those receiving placebo on both day 5 and day 10 [42]. Dodd also reported the protective efficacy of sucralfate mouthwash (every 6 h) among patients with gastrointestinal malignancies with 5-FU-based chemotherapy-induced OM [42].

However, many other RCTs have reported the ineffectiveness of sucralfate for treating and preventing OM induced by head and neck radiotherapy [43–45] and 5-FU-based chemotherapy [46, 47]. Therefore, the protective and treatment effects of sucralfate on cancer treatment-induced OM are inconsistent, and clinical guidelines do not currently recommend sucralfate use [48].

4.3.4 Glutamine

Glutamine is a precursor of glutathione, a major intracellular antioxidant that protects cells against oxidative stress, and is thought to protect the oral mucosal epithelium from oxidative stress by restoring tissue glutathione levels that are depleted after chemotherapy [49].

Two RCTs of patients receiving head and neck radiotherapy have reported the prophylactic effects of oral glutamine [50, 51]. Tsujimoto et al. examined 40 patients receiving head and neck chemoradiotherapy and reported that prophylactic oral glutamine administration (10 g, three times a day) significantly decreased the average grade of OM compared to placebo (2.9 ± 0.3 vs 3.3 ± 0.4, $P = 0.0049$) [50]. In an RCT of 70 patients receiving head and neck radiotherapy, Chattopadhyay et al. showed that patients who received 10 g of oral glutamine 2 h before radiotherapy had a significantly lower incidence of grade 3 and grade 4 OM than patients who did not ($P = 0.02$ and $P = 0.04$, respectively) [51]. Moreover, a metaanalysis assessing the effectiveness of glutamine for alleviating severe radiotherapy-induced OM in patients receiving head and neck radiotherapy examined 5 clinical studies comprising 234 patients with head and

Management of cancer treatment-induced oral mucositis **Chapter | 10 189**

neck cancer and showed a statistically significant benefit in reducing the risk and severity of radiotherapy-induced OM compared to placebo and no treatment (risk ratio 0.17, 95% CI 0.06–0.47) [52].

Furthermore, in an RCT of 24 cancer patients receiving chemotherapy, Anderson et al. reported that glutamine significantly reduced the duration and severity of OM compared to placebo ($P=0.0005$ and $P=0.002$, respectively) [53]. A systematic review of 15 studies in patients with various types of cancer further showed that oral glutamine significantly reduced the incidence of grade 2, 3, and 4 chemotherapy- and/or radiotherapy-induced OM [53].

However, in a randomized phase II study examining the efficacy and safety of glutamine and an elemental diet, Tanaka et al. found that while glutamine plus an elemental diet showed significant preventive effects on the severity of OM, glutamine alone showed no effect [54]. Moreover, in an RCT examining the effectiveness of parenteral glutamine supplementation in 40 autologous transplant patients, Pytlik et al. reported that patients who received parenteral glutamine had more severe OM than patients who received an intravenous glutamine-free amino acid solution ($P=0.04$); in addition, more relapses ($P=0.02$) and deaths ($P=0.05$) were observed among those receiving glutamine [55].

4.3.5 Zinc sulfate

A prospective, randomized, placebo-controlled study of 24 patients with head and neck cancer receiving radiotherapy by Ertekin et al. revealed that patients receiving zinc sulfate had significantly lower severity of mucositis and pharyngitis compared to those receiving placebo ($P<0.05$) [56]. However, in a randomized double-blind, placebo-controlled trial of zinc sulfate supplementation for radiation-induced OM in 140 patients with head and neck cancer, Sangthawan et al. reported that there were no significant differences in the frequency of grade ≥ 2 OM or pharyngitis at each week of radiation and at the first month after completion of radiation between zinc sulfate and placebo-treated patients [57]. Similarly, a double-blind, randomized, placebo-controlled study in 60 patients undergoing HSCT by Mansouri et al. reported that treatment with zinc sulfate had no significant effect on the development of chemotherapy-induced OM compared to placebo [58]. Accordingly, the protective and treatment effects of zinc sulfate on cancer treatment-induced OM remain unclear.

4.3.6 Benzydamine

Benzydamine (1-benzyl-3-(3-dimethylamino) propoxy-1H-indazole) is a nonsteroidal antiinflammatory drug (NSAID) that has been shown to exert antioxidant and antiinflammatory effects by decreasing the synthesis of IL-1β and TNF-α [59,60].

Several studies have reported the positive effects of benzydamine oral rinse in preventing radiation-induced OM [61–64]. First, a double-blind, randomized clinical study examining the effectiveness of benzydamine in 67 patients with radiation-induced mucositis of the oropharynx by Kim et al. showed that

190 Inflammation and oral cancer

benzydamine provided significant relief from mouth and throat pain induced by radiotherapy [61]. Second, in a multicenter, randomized, double-blind, placebo-controlled clinical trial of 165 head and neck cancer patients receiving radiotherapy, Epstein et al. found that benzydamine oral rinse significantly reduced erythema and ulceration, by approximately 30% compared to placebo ($P=0.006$). Benzydamine also significantly delayed the use of systemic analgesics ($P<0.05$) [62]. Third, Kazemian et al. demonstrated in a double-blind placebo controlled randomized trial of 81 head and neck cancer patients receiving radiotherapy that those who were also given benzydamine had significantly lower incidence of grade ≥ 3 OM than those given a placebo (43.6% vs 78.6%, $P=0.001$) [63]. Finally, a double-blind placebo-controlled randomized clinical trial of 51 patients with head and neck carcinoma receiving radiotherapy reported by Sheibani et al. showed that those receiving placebo had a higher mean mucositis score than those receiving benzydamine by the end of week 4 (1.81 vs 1.27, $P=0.001$), and that this trend continued until the end of week 7 (1.98 vs 1.43, $P=0.001$) [64].

Thus, benzydamine oral rinse is recommended for the prevention of radiotherapy-induced mucositis [65].

4.3.7 Palifermin

Palifermin, a recombinant human keratinocyte growth factor-1, is an approved agent for preventing severe OM induced by high-dose chemotherapy and radiotherapy followed by HSCT [66]. Several studies have reported the prophylactic effects of palifermin against OM in patients undergoing HSCT [67–69]. In a placebo-controlled, double-blind, phase 3 study, Spielberger et al. compared the effect of palifermin with that of a placebo on the development of OM in 212 patients with hematologic cancers who were undergoing autologous HSCT after receiving total-body irradiation and high-dose chemotherapy. They demonstrated that those receiving palifermin had a significantly lower incidence of grade 3 or 4 OM, based on the World Health Organization's criteria, compared to those receiving placebo (63% vs 98%, $P<0.001$) [67]. Moreover, palifermin was associated with a significantly shorter median duration of grade 3 or 4 mucositis (6 days (range 0–22) vs 9 days (range 0–27), $P<0.001$), reduced patient-reported soreness of the mouth and throat (area-under-the-curve score, 29.0 (range 0–98) vs 46.8 (range 0–110), $P<0.001$), and a reduction in the use of opioid analgesics (median, 212 mg of morphine equivalents (range 0–9418) vs 535 mg of morphine equivalents (range 0–9418), $P<0.001$) and the incidence of use of total parenteral nutrition (31% vs 55%, $P<0.001$) compared to placebo. Lucchese et al. conducted a randomized-controlled trial to examine the efficacy of palifermin as a primary prophylaxis in pediatric patients with acute lymphoblastic leukemia undergoing HSCT and showed that palifermin use significantly reduced the incidence of grade 3 and 4 OM compared to control [68]. Moreover, Blazar et al. conducted a phase 1/2 randomized, placebo-control trial to examine the potential benefits of palifermin on acute graft-versus-host

Management of cancer treatment-induced oral mucositis **Chapter | 10** **191**

disease (GVHD) and hematopoietic recovery in allogeneic HSCT recipients and concluded that palifermin was generally safe in allogeneic HSCT but had no significant effect on engraftment, acute GVHD, or survival [69].

In contrast, the prophylactic effects of palifermin in head and neck cancer patients with radiotherapy-induced OM are controversial. In a randomized, placebo-controlled, double-blind study of head and neck cancer patients receiving chemoradiotherapy (palifermin arm: $n=94$, placebo arm: $n=94$), Le et al. reported that palifermin significantly reduced the incidence of severe OM (WHO grade 3 or 4) compared to placebo (54% vs 69%, $P=0.041$) [70]. Similarly, Henke et al. conducted a multicenter, double-blind, randomized, placebo-controlled trial in 186 patients with stages II to IVB carcinoma of the oral cavity, oropharynx, hypopharynx, or larynx and reported that palifermin significantly reduced the incidence of severe OM (WHO grade 3 or 4) (51% vs 67%, $P=0.027$) [71]. However, after multiplicity adjustment, the secondary efficacy endpoints in these RCTs, including median time to severe OM, median duration of severe OM, and incidence of xerostomia grade ≥ 2, were not statistically significantly different between the two arms. Additionally, average MTS (patients' reported mouth and throat soreness) score was not significantly different between the two arms with/without multiplicity adjustment. Moreover, in a multicenter, double-blind, randomized, placebo-controlled study of 99 head and neck cancer patients receiving chemoradiotherapy, Brizel et al. reported that palifermin had no significant effect on the median duration or incidence of grade 2 mucositis compared to placebo [72].

5. Effect of polaprezinc on chemotherapy- and/or radiotherapy-induced oral mucositis

Polaprezinc, a zinc-L-carnosine, is a zinc-containing molecule used as a therapy for gastric ulcers [73]. A multiinstitutional randomized controlled trial of 91 patients receiving chemotherapy followed by HSCT revealed that patients who started polaprezinc before chemotherapy (prevention group) showed a significantly reduced incidence of grade ≥ 2 but not grade ≥ 3 OM compared to those who started polaprezinc after developing grade 2 OM (control group) (44.7% vs 22.0%, $P=0.025$) [74]. There was no significant difference in the rate of engraftment (95.6% vs 97.2%, $P=0.693$) between the two groups. Several retrospective studies have also reported that polaprezinc significantly prevents the incidence of severe OM in patients undergoing HSCT [75–78].

Furthermore, a number of studies have reported the prophylactic effects of polaprezinc on radiotherapy-induced OM in patients with head and neck cancer. Watanabe et al. reported that head and neck cancer patients receiving radiochemotherapy who also received a polaprezinc oral rinse had significantly lower incidence rates of grade ≥ 3 OM, grade ≥ 3 oral pain, grade ≥ 2 xerostomia, and grade ≥ 3 taste disturbance compared to those who received an azulene oral rinse [79]. Moreover, in a retrospective analysis of 104 patients receiving

192   Inflammation and oral cancer

radiotherapy with a total radiation dose of 70 Gy for head and neck cancer, Suzuki et al. reported that polaprezinc significantly reduced the median duration of radiotherapy (HR 0.557, 95% CI 0.357–0.871, $P = 0.0149$) and median time to discharge after completion of radiotherapy (HR 0.604, 95% CI 0.386–0.946, $P = 0.028$) [80].

The precise mechanisms underlying the prophylactic effects of polaprezinc on OM remain to be clarified. Evidence suggests that this compound inhibits the production of inflammatory cytokines, including TNF-α, IL-1β, IL-6, and MMP-2, and induces antiinflammatory effects and immune responses via inhibition of NF-κB [81–84]. Additionally, polaprezinc inhibits the production of ROS and induces insulin-like growth factor-1 (IGF-1), a polypeptide that has an important role in gastric epithelial wound repair [85]. These actions are thought to contribute to polaprezinc's preventive effects on cancer chemotherapy- and/or radiotherapy-induced OM.

References

[1]   McGuire DB, Altomonte V, Peterson DE, Wingard JR, Jones RJ, Grochow LB. Patterns of mucositis and pain in patients receiving preparative chemotherapy and bone marrow transplantation. Oncol Nurs Forum 1993;20(10):1493–502.

[2]   Duncan GG, Epstein JB, Tu D, El Sayed S, Bezjak A, Ottaway J, Pater J, National Cancer Institute of Canada Clinical Trials Group. Quality of life, mucositis, and xerostomia from radiotherapy for head and neck cancers: a report from the NCIC CTG HN2 randomized trial of an antimicrobial lozenge to prevent mucositis. Head Neck 2005;27(5):421–8.

[3]   Elting LS, Cooksley C, Chambers M, Cantor SB, Manzullo E, Rubenstein EB. The burdens of cancer therapy. Clinical and economic outcomes of chemotherapy-induced mucositis. Cancer 2003;98(7):1531–9.

[4]   Elting LS, Cooksley CD, Chambers MS, Garden AS. Risk, outcomes, and costs of radiation-induced oral mucositis among patients with head-and-neck malignancies. Int J Radiat Oncol Biol Phys 2007;68(4):1110–20.

[5]   Murphy BA. Clinical and economic consequences of mucositis induced by chemotherapy and/or radiation therapy. J Support Oncol 2007;5(9 Suppl 4):13–21.

[6]   Trotti A, Bellm LA, Epstein JB, Frame D, Fuchs HJ, Gwede CK, Komaroff E, Nalysnyk L, Zilberberg MD. Mucositis incidence, severity and associated outcomes in patients with head and neck cancer receiving radiotherapy with or without chemotherapy: a systematic literature review. Radiother Oncol 2003;66(3):253–62.

[7]   Pico JL, Avila-Garavito A, Naccache P. Mucositis: its occurrence, consequences, and treatment in the oncology setting. Oncologist 1998;3(6):446–51.

[8]   Lalla RV, Bowen J, Barasch A, Elting L, Epstein J, Keefe DM, McGuire DB, Migliorati C, Nicolatou-Galitis O, Peterson DE, Raber-Durlacher JE, Sonis ST, Elad S. Mucositis Guidelines Leadership Group of the Multinational Association of Supportive Care in Cancer and International Society of Oral Oncology (MASCC/ISOO). MASCC/ISOO clinical practice guidelines for the management of mucositis secondary to cancer therapy. Cancer 2014;120(10):1453–61.

[9]   Kashiwazaki H, Matsushita T, Sugita J, Shigematsu A, Kasashi K, Yamazaki Y, Kanehira T, Yamamoto S, Kondo T, Endo T, Tanaka J, Hashino S, Nishio N, Imamura M, Kitagawa Y, Inoue N. Professional oral health care reduces oral mucositis and febrile neutropenia in patients treated with allogeneic bone marrow transplantation. Support Care Cancer 2012;20(2):367–73.

Management of cancer treatment-induced oral mucositis **Chapter | 10 193**

[10] Vigarios E, Epstein JB, Sibaud V. Oral mucosal changes induced by anticancer targeted therapies and immune checkpoint inhibitors. Support Care Cancer 2017;25(5):1713–39.

[11] Lalla RV, Sonis ST, Peterson DE. Management of oral mucositis in patients with cancer. Dent Clin N Am 2008;52(1):61–77.

[12] Sonis ST. New thoughts on the initiation of mucositis. Oral Dis 2010;16(7):597–600.

[13] Logan RM, Stringer AM, Bowen JM, Gibson RJ, Sonis ST, Keefe DM. Is the pathobiology of chemotherapy-induced alimentary tract mucositis influenced by the type of mucotoxic drug administered? Cancer Chemother Pharmacol 2009;63(2):239–51.

[14] Moura JF, Mota JM, Leite CA, Wong DV, Bezerra NP, Brito GA, Lima V, Cunha FQ, Ribeiro RA. A novel model of megavoltage radiation-induced oral mucositis in hamsters: role of inflammatory cytokines and nitric oxide. Int J Radiat Biol 2015;91(6):500–9.

[15] Min CK, Lee WY, Min DJ, Lee DG, Kim YJ, Park YH, Kim HJ, Lee S, Kim DW, Lee JW, Min WS, Kim CC. The kinetics of circulating cytokines including IL-6, TNF-a, IL-8 and IL-10 following allogeneic hematopoietic stem cell transplantation. Bone Marrow Transplant 2001;28(10):935–40.

[16] Silva GB, Sacono NT, Othon-Leite AF, Mendonça EF, Arantes AM, Bariani C, Duarte LG, Abreu MH, Queiroz-Júnior CM, Silva TA, Batista AC. Effect of low-level laser therapy on inflammatory mediator release during chemotherapy-induced oral mucositis: a randomized preliminary study. Lasers Med Sci 2015;30(1):117–26.

[17] Haddad R, Sonis S, Posner M, Wirth L, Costello R, Braschayko P, Allen A, Mahadevan A, Flynn J, Burke E, Li Y, Tishler RB. Randomized phase 2 study of concomitant chemoradiotherapy using weekly carboplatin/paclitaxel with or without daily subcutaneous amifostine in patients with locally advanced head and neck cancer. Cancer 2009;115(19):4514–23.

[18] Yeo E, Alvarado T, Fainstein V, Bodey GP. Prophylaxis of oropharyngeal candidiasis with clotrimazole. J Clin Oncol 1985;3(12):1668–71.

[19] Greenberg MS, Friedman H, Cohen SG, Oh SH, Laster L, Starr S. A comparative study of herpes simplex infections in renal transplant and leukemic patients. J Infect Dis 1987;156(2):280–7.

[20] Sonis ST. Pathobiology of mucositis. Semin Oncol Nurs 2004;20(1):11–5.

[21] Hong CHL, Gueiros LA, Fulton JS, Cheng KKF, Kandwal A, Galiti D, Fall-Dickson JM, Johansen J, Ameringer S, Kataoka T, Weikel D, Eilers J, Ranna V, Vaddi A, Lalla RV, Bossi P, Elad S. Mucositis Study Group of the Multinational Association of Supportive Care in Cancer/International Society for Oral Oncology (MASCC/ISOO). Systematic review of basic oral care for the management of oral mucositis in cancer patients and clinical practice guideline. Support Care Cancer 2019;27(10):3949–67.

[22] Mahood DJ, Dose AM, Loprinzi CL, Veeder MH, Athmann LM, Therneau TM, Sorensen JM, Gainey DK, Mailliard JA, Gusa NL. Inhibition of fluorouracil-induced stomatitis by oral cryotherapy. J Clin Oncol 1991;9(3):449–52.

[23] Walladbegi J, Gellerstedt M, Svanberg A, Jontell M. Innovative intraoral cooling device better tolerated and equally effective as ice cooling. Cancer Chemother Pharmacol 2017;80(5):965–72.

[24] Correa MEP, Cheng KKF, Chiang K, Kandwal A, Loprinzi CL, Mori T, Potting C, Rouleau T, Toro JJ, Ranna V, Vaddi A, Peterson DE, Bossi P, Lalla RV, Elad S. Systematic review of oral cryotherapy for the management of oral mucositis in cancer patients and clinical practice guidelines. Support Care Cancer 2020;28(5):2449–56.

[25] Regan V, Servinova E, Packer L. Antioxidant effects of ubiquinones in microsomes and mitochondria are mediated by tocopherol recycling. Biochem Biophys Res Commun 1990;169(3):851–6.

194 Inflammation and oral cancer

[26] Tampo Y, Yonaha M. Vitamin E and gluthatione are required for preservation of microsomal gluthatione S-transferase from oxidative stress in microsomes. Pharmacology 1990;66(4):259–65.

[27] Starasoler S, Haber GS. Use of vitamin E oil in primary herpes gingivostomatitis in an adult. NY State Dentristy 1978;44:382–3.

[28] Wadleigh RG, Redman RS, Graham ML, Krasnow SH, Anderson A, Cohen MH. Vitamin E in the treatment of chemotherapy-induced mucositis. Am J Med 1992;92(5):481–4.

[29] Khurana H, Pandey RK, Saksena AK, Kumar A. An evaluation of vitamin E and Pycnogenol in children suffering from oral mucositis during cancer chemotherapy. Oral Dis 2013;19(5):456–64.

[30] Ferreira PR, Fleck JF, Diehl A, Barletta D, Braga-Filho A, Barletta A, Ilha L. Protective effect of alpha-tocopherol in head and neck cancer radiation-induced mucositis: a doubleblind randomized trial. Head Neck 2004;26(4):313–21.

[31] Sung L, Tomlinson GA, Greenberg ML, Koren G, Judd P, Ota S, Feldman BM. Serial controlled N-of-1 trials of topical vitamin E as prophylaxis for chemotherapy-induced oral mucositis in paediatric patients. Eur J Cancer 2007;43(8):1269–75.

[32] El-Housseiny AA, Saleh SM, El-Masry AA, Allam AA. The effectiveness of vitamin "E" in the treatment of oral mucositis in children receiving chemotherapy. J Clin Pediatr Dent 2007;31(3):167–70.

[33] Nakamura K, Natsugoe S, Kumanohoso T, Shinkawa T, Kariyazono H, Yamada K, Baba M, Yoshinaka H, Fukumoto T, Aikou T. Prophylactic action of allopurinol against chemotherapy-induced stomatitis—inhibition of superoxide dismutase and proteases. Anticancer Drugs 1996;7(3):235–9.

[34] Porta C, Moroni M, Nastasi G. Allopurinol mouthwashes in the treatment of 5-fluorouracilinduced stomatitis. Am J Clin Oncol 1994;17(3):246–7.

[35] Yokomizo H, Yoshimatsu K, Hashimoto M, Ishibashi K, Umehara A, Yoshida K, Fujimoto T, Watanabe K, Ogawa K. Prophylactic efficacy of allopurinol ice ball for leucovorin/5-fluorouracil therapy-induced stomatitis. Anticancer Res 2004;24(2C):1131–4.

[36] Elzawawy A. Treatment of 5-fluorouracil-induced stomatitis by allopurinol mouthwashes. Oncology 1991;48(4):282–4.

[37] Panahi Y, Ala S, Saeedi M, Okhovatian A, Bazzaz N, Naghizadeh MM. Allopurinol mouth rinse for prophylaxis of fluorouracil-induced mucositis. Eur J Cancer Care 2010;19(3):308–12.

[38] Weiss GR, Green S, Hannigan EV, Boutselis JG, Surwit EA, Wallace DL, Alberts DS. A phase II trial of cisplatin and 5-fluorouracil with allopurinol for recurrent or metastatic carcinoma of the uterine cervix: a southwest oncology group trial. Gynecol Oncol 1990;37(3):354–8.

[39] Szabo S, Hollander D. Pathways of gastrointestinal protection and repair: mechanisms of action of sucralfate. Am J Med 1989;86(6 Suppl 1):23–31. McCarthy DM. Sucralfate. N Engl J Med 1991;325(14):1017–1025.

[40] Etiz D, Erkal HS, Serin M, Küçük B, Hepari A, Elhan AH, Tulunay O, Cakmak A. Clinical and histopathological evaluation of sucralfate in prevention of oral mucositis induced by radiation therapy in patients with head and neck malignancies. Oral Oncol 2000;36(1):116–20.

[41] Castagna L, Benhamou E, Pedraza E, Luboinski M, Forni M, Brandes I, Pico JL, Dietrich PY. Prevention of mucositis in bone marrow transplantation: a double blind randomised controlled trial of sucralfate. Ann Oncol 2001;12(7):953–5.

[42] Ala S, Saeedi M, Janbabai G, Ganji R, Azhdari E, Shiva A. Efficacy of sucralfate mouth wash in prevention of 5-fluorouracil induced oral mucositis: a prospective, randomized, double-blind. Controlled Trial Nutr Cancer 2016;68(3):456–63.

Management of cancer treatment-induced oral mucositis **Chapter | 10** **195**

[43] Carter DL, Hebert ME, Smink K, Leopold KA, Clough RL, Brizel DM. Double blind randomized trial of sucralfate vs placebo during radical radiotherapy for head and neck cancers. Head Neck 1999;21(8):760–6.

[44] Makkonen TA, Boström P, Vilja P, Joensuu H. Sucralfate mouth washing in the prevention of radiation-induced mucositis: a placebo-controlled double-blind randomized study. Int J Radiat Oncol Biol Phys 1994;30(1):177–82.

[45] Epstein JB, Wong FL. The efficacy of sucralfate suspension in the prevention of oral mucositis due to radiation therapy. Int J Radiat Oncol Biol Phys 1994;28(3):693–8.

[46] Loprinzi CL, Ghosh C, Camoriano J, Sloan J, Steen PD, Michalak JC, Schaefer PL, Novotny PJ, Gerstner JB, White DF, Hatfield AK, Quella SK. Phase III controlled evaluation of sucralfate to alleviate stomatitis in patients receiving fluorouracil-based chemotherapy. J Clin Oncol 1997;15(3):1235–8.

[47] Nottage M, McLachlan SA, Brittain MA, Oza A, Hedley D, Feld R, Siu LL, Pond G, Moore MJ. Sucralfate mouthwash for prevention and treatment of 5-fluorouracil-induced mucositis: a randomized, placebo-controlled trial. Support Care Cancer 2003;11(1):41–7.

[48] Saunders DP, Rouleau T, Cheng K, Yarom N, Kandwal A, Joy J, Bektas Kayhan K, van de Wetering M, Brito-Dellan N, Kataoka T, Chiang K, Ranna V, Vaddi A, Epstein J, Lalla RV, Bossi P, Elad S. Mucositis Study Group of the Multinational Association of Supportive Care in Cancer/International Society of Oral Oncology (MASCC/ISOO). Systematic review of antimicrobials, mucosal coating agents, anesthetics, and analgesics for the management of oral mucositis in cancer patients and clinical practice guidelines. Support Care Cancer 2020;28(5):2473–84.

[49] Leitao RF, Ribeiro RA, Lira AM, Silva LR, Bellaguarda EA, Macedo FD, Sousa RB, Brito GAC. Glutamine and alanyl-glutamine accelerate the recovery from 5-fluorouracil-induced experimental oral mucositis in hamster. Cancer Chemother Pharmacol 2008;61(2):215–22.

[50] Tsujimoto T, Yamamoto Y, Wasa M, Takenaka Y, Nakahara S, Takagi T, Tsugane M, Hayashi N, Maeda K, Inohara H, Uejima E, Ito T. L-glutamine decreases the severity of mucositis induced by chemoradiotherapy in patients with locally advanced head and neck cancer: a doubleblind, randomized, placebo-controlled trial. Oncol Rep 2015;33(1):33–9.

[51] Chattopadhyay S, Saha A, Azam M, Mukherjee A, Sur PK. Role of oral glutamine in alleviation and prevention of radiation-induced oral mucositis: a prospective randomized study. South Asian J Cancer 2014;3(1):8–12.

[52] Leung HW, Chan AL. Glutamine in alleviation of radiation-induced severe oral mucositis: a meta-analysis. Nutr Cancer 2016;68(5):734–42.

[53] Anderson PM, Schroeder G, Skubitz KM. Oral glutamine reduces the duration and severity of stomatitis after cytotoxic cancer chemotherapy. Cancer 1998;83(7):1433–9.

[54] Sayles C, Hickerson SC, Bhat RR, Hall J, Garey KW, Trivedi MV. Oral glutamine in preventing treatment-related mucositis in adult patients with cancer: a systematic review. Nutr Clin Pract 2016;31(2):171–9.

[55] Pytlik R, Benes P, Patorková M, Chocenská E, Gregora E, Procházka B, Kozák T. Standardized parenteralalanyl-glutamine dipeptide supplementation is not beneficial in autologous transplant patients: a randomized, double-blind, placebo controlled study. Bone Marrow Transplant 2002;30(12):953–61.

[56] Ertekin MV, Koç M, Karslioglu I, Sezen O. Zinc sulfate in the prevention of radiationinduced oropharyngeal mucositis: a prospective, placebo-controlled, randomized study. Int J Radiat Oncol Biol Phys 2004;58(1):167–74.

[57] Sangthawan D, Phungrassami T, Sinkitjarurnchai W. A randomized double-blind, placebo-controlled trial of zinc sulfate supplementation for alleviation of radiationinduced oral mucositis and pharyngitis in head and neck cancer patients. J Med Assoc Thai 2013;96(1):69–76.

196 Inflammation and oral cancer

[58] Mansouri A, Hadjibabaie M, Iravani M, Shamshiri AR, Hayatshahi A, Javadi MR, Khoee SH, Alimoghaddam K, Ghavamzadeh A. The effect of zinc sulfate in the prevention of highdose chemotherapy-induced mucositis: a doubleblind, randomized, placebo-controlled study. Hematol Oncol 2012;30(1):22–6.

[59] Quane PA, Graham GG, Ziegler JB. Pharmacology of benzydamine. Inflammopharmacology 1998;6(2):95–107.

[60] Sironi M, Milanese C, Vecchi A, Polenzani L, Guglielmotti A. Coletta benzydamine inhibits the release of tumor necrosis factor-alpha and monocyte chemotactic protein-1 by *Candida albicans*-stimulated human peripheral blood cells. Int J Clin Lab Res 1997;27(2):112–8.

[61] Kim JH, Chu F, Lakshmi V, Houde R. A clinical study of benzydamine for the treatment of radiotherapy-induced mucositis of the oropharynx. Int J Tissue React 1985;7(3):215–8.

[62] Epstein JB, Silverman Jr S, Paggiarino DA, Crockett S, Schubert MM, Senzer NN, Lockhart PB, Gallagher MJ, Peterson DE, Leveque FG. Benzydamine HCl for prophylaxis of radiation induced oral mucositis: results from a multicenter, randomized, double-blind, placebo-controlled clinical trial. Cancer 2001;92(4):875–85.

[63] Kazemian A, Kamian S, Aghili M, Hashemi FA, Haddad P. Benzydamine for prophylaxis of radiation-induced oral mucositis in head and neck cancers: a double-blind placebo-controlled randomized clinical trial. Eur J Cancer Care 2009;18(2):174–8.

[64] Sheibani KM, Mafi AR, Moghaddam S, Taslimi F, Amiran A, Ameri A. Efficacy of benzydamine oral rinse in prevention and management of radiation-induced oral mucositis: a double-blind placebo-controlled randomized clinical trial. Asia Pac J Clin Oncol 2015;11:22–7.

[65] Ariyawardana A, Cheng KKF, Kandwal A, Tilly V, Al-Azri AR, Galiti D, Chiang K, Vaddi A, Ranna V, Nicolatou-Galitis O, Lalla RV, Bossi P, Elad S. Mucositis Study Group of the Multinational Association of Supportive Care in Cancer/International Society for Oral Oncology (MASCC/ISOO). Systematic review of anti-inflammatory agents for the management of oral mucositis in cancer patients and clinical practice guidelines. Support Care Cancer 2019;27(10):3985–95.

[66] Finch PW, Rubin JS. Keratinocyte growth factor/fibroblast growth factor 7, a homeostatic factor with therapeutic potential for epithelial protection and repair. Adv Cancer Res 2004;91:69–136.

[67] Spielberger R, Stiff P, Bensinger W, Gentile T, Weisdorf D, Kewalramani T, Shea T, Yanovich S, Hansen K, Noga S, McCarty J, LeMaistre CF, Sung EC, Blazar BR, Elhardt D, Chen MG, Emmanouilides C. Palifermin for oral mucositis after intensive therapy for hematologic cancers. N Engl J Med 2004;351(25):2590–8.

[68] Lucchese A, Matarese G, Ghislanzoni LH, Gastaldi G, Manuelli M, Gherlone E. Efficacy and effects of palifermin for the treatment of oral mucositis in patients affected by acute lymphoblastic leukemia. Leuk Lymphoma 2016;57(4):820–7.

[69] Blazar BR, Weisdorf DJ, Defor T, Goldman A, Braun T, Silver S, Ferrara JL. Phase 1/2 randomized, placebo-control trial of palifermin to prevent graft-versus-host disease (GVHD) after allogeneic hematopoietic stem cell transplantation (HSCT). Blood 2006;108(9):3216–22.

[70] Le QT, Kim HE, Schneider CJ, Muraközy G, Skladowski K, Reinisch S, Chen Y, Hickey M, Mo M, Chen MG, Berger D, Lizambri R, Henke M. Palifermin reduces severe mucositis in definitive chemoradiotherapy of locally advanced head and neck cancer: a randomized, placebocontrolled study. J Clin Oncol 2011;29:2808–14.

[71] Henke M, Alfonsi M, Foa P, Giralt J, Bardet E, Cerezo L, Salzwimmer M, Lizambri R, Emmerson L, Chen MG, Berger D. Palifermin decreases severe oral mucositis of patients undergoing postoperative radiochemotherapy for head and neck cancer: a randomized, placebocontrolled trial. J Clin Oncol 2011;29(20):2815–20.

Management of cancer treatment-induced oral mucositis **Chapter | 10 197**

[72] Brizel DM, Murphy BA, Rosenthal DI, Pandya KJ, Glück S, Brizel HE, Meredith RF, Berger D, Chen MG, Mendenhall W. Phase II study of palifermin and concurrent chemoradiation in head and neck squamous cell carcinoma. J Clin Oncol 2008;26(15):2489–96.

[73] Dajani EZ, Klamut MJ. Novel therapeutic approaches to gastric and duodenal ulcers: an update. Expert Opin Investig Drugs 2000;9(7):1537–44.

[74] Kitagawa J, Kobayashi R, Nagata Y, Kasahara S, Ono T, Sawada MD, Ohata K, Kato-Hayashi H, Hayashi H, Shimizu M, Itoh Y, Tsurumi H, Suzuki A. Polaprezinc for prevention of oral mucositis in patients receiving chemotherapy followed by hematopoietic stem cell transplantation: a multi-institutional randomized controlled trial. Int J Cancer 2021;148(6):1462–9.

[75] Tsubura-Okubo M, Komiyama Y, Kamimura R, Sawatani Y, Arai H, Mitani K, Haruyama Y, Kobashi G, Ishihama H, Uchida D, Kawamata H. Oral management with polaprezinc solution reduces adverse events in haematopoietic stem cell transplantation patients. Int J Oral Maxillofac Surg 2020. S0901-5027(20)30382–30389.

[76] Funato M, Ozeki M, Suzuki A, Ishihara M, Kobayashi R, Fukao T, Ioth Y. Prophylactic effect of polaprezinc, a zinc-L-carnosine, against chemotherapy-induced oral mucositis in pediatric patients undergoing autologous stem cell transplantation. Anticancer Res 2018;38(8):4691–7.

[77] Hayashi H, Kobayashi R, Suzuki A, Yamada Y, Ishida M, Shakui T, Kitagawa J, Hayashi H, Sugiyama T, Takeuchi H, Tsurumi H, Itoh Y. Preparation and clinical evaluation of a novel lozenge containing polaprezinc, a zinc-L-carnosine, for prevention of oral mucositis in patients with hematological cancer who received high-dose chemotherapy. Med Oncol 2016;33(8):91.

[78] Hayashi H, Kobayashi R, Suzuki A, Ishihara M, Nakamura N, Kitagawa J, Kanemura N, Kasahara S, Kitaichi K, Hara T, Tsurumi H, Moriwaki H, Itoh Y. Polaprezinc prevents oral mucositis in patients treated with high dose chemotherapy followed by hematopoietic stem cell transplantation. Anticancer Res 2014;34(12):7271–7.

[79] Watanabe T, Ishihara M, Matsuura K, Mizuta K, Itoh Y. Polaprezinc prevents oral mucositis associated with radiochemotherapy in patients with head and neck cancer. Int J Cancer 2010;127:1984–90.

[80] Suzuki A, Kobayashi R, Shakui T, Kubota Y, Fukita M, Kuze B, Aoki M, Sugiyama T, Mizuta K, Itoh Y. Effect of polaprezinc on oral mucositis, irradiation period, and time to discharge in patients with head and neck cancer. Head Neck 2016;38(9):1387–92.

[81] Ko JK, Leung CC. Ginger extract and polaprezinc exert gastroprotective actions by anti-oxidant and growth factor modulating effects in rats. J Gastroenterol Hepatol 2010;25(12):1861–8.

[82] Ueda K, Ueyama T, Oka M, Ito T, Tsuruo Y, Ichinose M. Polaprezinc (zinc L-carnosine) is a potent inducer of anti-oxidative stress enzyme, heme oxygenase (HO)-1—a new mechanism of gastric mucosal protection. J Pharmacol Sci 2009;110(3):285–94.

[83] Naito Y, Yoshikawa T, Yagi N, Matsuyama K, Yoshida N, Seto K, Yoneta T. Effects of polaprezinc on lipid peroxidation, neutrophil accumulation, and TNF-alpha expression in rats with aspirin-induced gastric mucosal injury. Dig Dis Sci 2001;46(4):845–51.

[84] Odashima M, Otaka M, Jin M, Wada I, Horikawa Y, Matsuhashi T, Ohba R, Hatakeyama N, Oyake J, Watanabe S. Zinc L-carnosine protects colonic mucosal injury through induction of heat shock protein 72 and suppression of NF-kappaB activation. Life Sci 2006;79(24):2245–50.

[85] Kato S, Tanaka A, Ogawa Y, Kanatsu K, Seto K, Yoneda T, Takeuchi K. Effect of polaprezinc on impaired healing of chronic gastric ulcers in adjuvant-induced arthritic rats—role of insulin-like growth factors (IGF)-1. Med Sci Monit 2001;7(1):20–5.

Chapter 11

Perspectives in research on oral squamous cell carcinoma

Hiroyuki Tomita
Department of Tumor Pathology, Gifu University Graduate School of Medicine, Gifu, Japan

Oral squamous cell carcinoma (OSCC) originates from the oral mucosal epithelial cells and commonly occurs in the larynx, pharynx, nasal cavity, or paranasal sinuses. OSCC comprises $\geq 90\%$ of oral cancer cases. In several studies, the term "oral cancer" has been used synonymously with "OSCC." With increasing information on the risk factors, gene mutations, epigenetic alterations, and clinical factors, there is a need for separate studies on tumors with different loci.

OSCC has one of the highest global incidence rates among cancers. Yearly, at least 350,000 new cases and approximately 180,000 deaths are registered worldwide. However, there are major geographical and environmental differences in risk factors [1]. The OSCC onset rate is currently decreasing in some countries; however, it is increasing in others, especially low-income countries and among women [2–4]. In addition, the OSCC occurrence rate in younger people (persons aged ≤ 45 years) is increasing. Approximately 80% of smokers worldwide are found in developing countries, and alcohol and tobacco consumption among females has increased; this might explain the increase in the OSCC onset rate. However, this does not explain the increase in the onset rate among younger patients, who have a shorter time of exposure to these risk factors. This is indicative of the necessity to tackle OSCC-related challenges in young people, including risk factors, the pattern of hereditary of causative genetic mutations, clinical behavior, and prognosis. Efficacious programs for stopping or reducing tobacco (smoked or chewed) and alcohol consumption would be highly effective in reducing the onset rates of OSCC and other cancers associated with these risk factors.

Oral potentially malignant disorders (OPMDs), such as leukoplakia, erythroplakia, oral mucosal hyperplasia, and proliferative verrucous leukoplakia (PVL), develop before OSCC. Among the OPMDs, PVL has unique characteristics: it is not always related to the classic environmental factors, the natural history is different from other OPMDs, and it has the highest probability of exacerbation among all OPMDs [5, 6]. Additionally, the status of oral lichen planus (OLP) as another OPMD remains uncertain; however, according to some

---

Inflammation and Oral Cancer. https://doi.org/10.1016/B978-0-323-88526-3.00011-7
Copyright © 2022 Elsevier Inc. All rights reserved.

200 Inflammation and oral cancer

recently metaanalyses, although the rate of progression from OLP to OSCC is low, OLP may be thought to be an OPMD [7–9]. Prompt histological evaluation and therapy of OPMD are very important for eliminating or extensively reducing the risk of malignant transformation. It has also been suggested that radical treatment is not appropriate for some lesions and that not all cases have malignant potential. In addition, the histological evaluation of epithelial dysplasia may be subjective, and according to certain metaanalyses, the mean overall rate of malignant transformation of OPMDs with associated dysplasia does not exceed 12.1% [10, 11]. Therefore, for rational treatment planning, proper follow-up, and planning of a highly cost-effective oral screening program, elucidating the characteristics of biomarkers that are indicative of the malignant risk magnitude, mechanisms of malignancy, and time before progression to malignancy is extremely important.

Non-OSCC neoplasms may develop in other sites, i.e., mesenchymal tissues, or other organs, within the oral cavity, or originate from distant metastases. Unlike OSCC, these have lower onset rates. Areas requiring further research include the mechanisms of pathogenesis and pathology, characteristics at diagnosis, treatment strategies, and prognostic markers.

OSCC is a highly invasive tumor, and most patients present with locally progressive lesions at diagnosis; therefore, multifaceted treatment is essential. The principal causes of death in patients with OSCC are tumor invasion, lymph node metastasis, local recurrence, and secondary primary cancers. However, tumors in the tongue and oral cavity are highly invasive even at early stages and have a high invasion and metastasis potential. Thus, the survival rate is only 40%–50%, which has not increased for decades [12, 13]. A thorough understanding of the mechanisms of tumor onset and progression has been achieved, and numerous biomarkers of OSCC diagnosis and prognosis have been proposed; however, no biomarker meets the rigorous criteria for use in clinical practice.

Experts in the field have agreed that to progress the management of OSCC patients, biomarkers that are clinically useful for early diagnosis, targeted treatment, treatment reaction, prognosis, and follow-up are emergently required. Ideally, various approaches should be considered. These include (i) a prospective analysis of multiple potential biomarkers in a large multicenter cohort study, if possible, to elucidate the characteristics of OSCC biomarkers and (ii) use of numerical and alternative methods in determining the effects of biomarkers. To date, most studies have been based solely on immunohistochemistry, without standardized quantitative profiles. In addition, recognizing the important strong and weak points of cancer heterogeneity in the identification of prognostic biomarkers is essential. Most cancers have a single cell type at onset; however, they tend to become heterogeneous (have multiple tumor cell types) as the cancer progresses. In addition, irradiation, which is likely to cause DNA damage, is considered a potential source of field cancerization of OSCCs. Radiotherapy for OSCCs has been examined as a possible risk factor for second primary cancers. Thus, OSCC development in the irradiation field is a factor favoring the tumor development among the irradiation field, which is a probable cause of clinical disease heterogeneity Even for the same disease, OSCC, differences in loci, clinical disease stage, and histopathological characteristics, might imply that the cells have different profiles of genetic and epigenetic alterations and contribute to clinical prognosis. While cancer stem cell markers are expected to be a useful biomarker and their therapeutic targets is promising, currently available data is insufficient.

The most important requirements for providing preclinical data to support clinical progress include identification of high-risk mutations, mutated genes, signaling pathways, and elucidation of the tumor microenvironment and its constituted cells. Using data from large-scale studies, various omic technologies, including genomics, epigenomics, transcriptomics, proteomics, and metabolomics, are essential tools for elucidating the functional roles of specific genes using the corresponding RNA (including regulatory factors such as noncoding RNA) and proteins, and identifying patients with different cancers who are eligible for treatment. Numerous genome-wide profiling studies have been performed using OSCC samples, and the results may be useful in the identification of biomarkers and targeted treatment, resulting in higher survival rates.

In addition, liquid biopsies have recently attracted considerable attention as potential tools for identifying cancer biomarkers by analyzing biological fluid samples, such as blood and saliva. Many useful candidates, i.e., DNA of circulating tumor cells and exosomes, in such readily accessible body fluids can thus be evaluated. These are potentially ideal methods for OSCC screening programs. Identifying the biomarkers in these fluids and designing clinical validation studies with long-term outcomes are required.

Relationships of oral cancers with periodontitis and other chronic inflammatory diseases have been suggested. Moreover, focus has been placed on the longstanding idea of the contribution of poor oral hygiene in the development of oral cancers. There are hundreds of bacterial species in the oral cavity, and the implication of one or several of these species in tumor onset and/or progression has been suggested. Several species of bacteria are associated with periodontal disease and oral caners, such as *Fusobacterium nucleatum*, *Prevotella intermedia*, and *Porphyromonas gingivalis* [14, 15], but the relationships between these mechanisms remain unclear. One possible mechanism is that oral bacteria trigger host cell intracellular signal transmission, with chemokine synthesis suppressing proliferation of host epithelial cells and regulating invasion of local microenvironments by immune cells. Therefore, a detailed understanding of the roles of oral microbiomes generated by oral cancers is an important research area, not only for oral cancer researchers, but also gastrointestinal tract researchers [16]. As suggested above, the findings of such research may be useful in explaining the higher than expected occurrence rate of oral cancer among patients without risk factors such as alcohol and tobacco consumption.

Surgery remains the mainstay of treatment for OSCC. Adjuvant radiotherapy is performed concomitantly with chemotherapy for advanced cases. However, recent studies have shown very favorable results with immunotherapy and targeted therapy. For example, treatment of recurrent and metastatic cancer with programmed death-1 (PD-1) and programmed death-ligand 1 (PD-L1) immune checkpoint inhibitors has been validated in a phase III, randomized, clinical study. It has recently been suggested that these agents can also be used to treat other refractory areas [12, 17]. The inhibitors of PD-1 are effective to improve results in clinical patients who are resistant to platinum-based chemotherapy and to prolong survival time. In addition, pembrolizumab, a humanized anti-PD-1 antibody, has been approved as a monotherapy for recurrent and/or metastatic cancer with a PD-1 ligand, PD-L1, overexpression.

A basis for planning and implementing standard treatment strategies for malignancy progression is provided by a global comprehension of the molecular incidents promoting tumorigenesis in the oral cavity [18]. In response to the identified epithelial growth factor receptor drive mechanism, a monoclonal antibody and low-molecular-weight inhibitor are being used [19, 20]. However, in many cases, there have been no definite clinical results, and there are major barriers to progress. Furthermore, tumor microenvironment components and cell groups may be potential biomarkers and treatment targets, which may be better alternatives to the currently used tumor-cell-focused treatment methods. According to specialists in clinical and fundamental research on oral cancers, the main areas for research are identification and overcoming of treatment-related challenges.

In addition, with the widespread use of immune checkpoint inhibitors, animal models are necessary in investigating the immune microenvironment of oral cancers. Human oral cancer cells only remain in place and proliferate in immunodeficient mice; therefore, there has been a delay in validating immune mechanisms [21]. However, the immune response should be verified using genetically modified mice or mice with chemically induced carcinogenesis. Only a few mouse models can be used to study the immune response associated with oral cancer development and the associated microenvironment; research is ongoing [22, 23]. To date, only a few histological images have been presented, and new models are indispensable. However, with the differences between human and murine immunity, there is progress in the development of mice with human-equivalent immunity.

The efficacy of oral cancer treatment methods is controversial, and treatment responses vary with patients. These differences may lead to major problems, such as unnecessary adverse effects and excessive treatment. Before anticancer drugs and radiation therapy can be applied to patients, it is necessary to obtain evidence to confirm their efficacy, and it is also important to establish a method to do so. Similar to other cancer types, using samples from patients with OSCC, some in vitro and in vivo disease models have been proposed for the validation of anticancer agents and radiotherapy; however, patients' responses have not been compared and verified directly with most of these models, and clinical application is therefore difficult. With the rapid advances in individualized

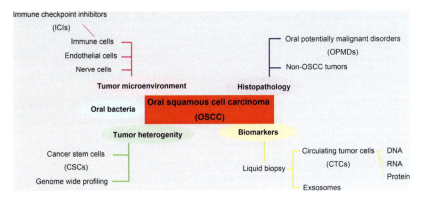

FIG. 1 Trends in basic and clinical research in 2021.

medical treatment, there is also a need to prioritize individualized management of OSCC, which is an important area for future research.

In conclusion, we need to be aware of the trends in basic and clinical research and work towards the control of oral cancer (Fig. 1).

References

[1] Bray F, Ferlay J, Soerjomataram I, Siegel RL, Torre LA, Jemal A. Global cancer statistics 2018: GLOBOCAN estimates of incidence and mortality worldwide for 36 cancers in 185 countries. CA Cancer J Clin 2018;68(6):394–424.

[2] Shield KD, Ferlay J, Jemal A, Sankaranarayanan R, Chaturvedi AK, Bray F, et al. The global incidence of lip, oral cavity, and pharyngeal cancers by subsite in 2012. CA Cancer J Clin 2017;67(1):51–64.

[3] Du M, Nair R, Jamieson L, Liu Z, Bi P. Incidence trends of lip, oral cavity, and pharyngeal cancers: global burden of disease 1990-2017. J Dent Res 2020;99(2):143–51.

[4] Miranda-Filho A, Bray F. Global patterns and trends in cancers of the lip, tongue and mouth. Oral Oncol 2020;102:104551.

[5] Thomson PJ, Goodson ML, Smith DR. Potentially malignant disorders revisited—the lichenoid lesion/proliferative verrucous leukoplakia conundrum. J Oral Pathol Med 2018;47(6):557–65.

[6] Iocca O, Sollecito TP, Alawi F, Weinstein GS, Newman JG, De Virgilio A, et al. Potentially malignant disorders of the oral cavity and oral dysplasia: a systematic review and meta-analysis of malignant transformation rate by subtype. Head Neck 2020;42(3):539–55.

[7] Aghbari SMH, Abushouk AI, Attia A, Elmaraezy A, Menshawy A, Ahmed MS, et al. Malignant transformation of oral lichen planus and oral lichenoid lesions: a meta-analysis of 20095 patient data. Oral Oncol 2017;68:92–102.

[8] Richards D. Malignant transformation rates in oral lichen planus. Evid Based Dent 2018;19(4):122.

[9] Giuliani M, Troiano G, Cordaro M, Corsalini M, Gioco G, Lo Muzio L, et al. Rate of malignant transformation of oral lichen planus: a systematic review. Oral Dis 2019;25(3):693–709.

[10] Mehanna HM, Rattay T, Smith J, McConkey CC. Treatment and follow-up of oral dysplasia—a systematic review and meta-analysis. Head Neck 2009;31(12):1600–9.

204 Inflammation and oral cancer

[11] Panwar A, Lindau R, Wieland A. Management for premalignant lesions of the oral cavity. Expert Rev Anticancer Ther 2014;14(3):349–57.

[12] Huber MA, Tantiwongkosi B. Oral and oropharyngeal cancer. Med Clin North Am 2014;98(6):1299–321.

[13] Chi AC, Day TA, Neville BW. Oral cavity and oropharyngeal squamous cell carcinoma—an update. CA Cancer J Clin 2015;65(5):401–21.

[14] Mager DL, Haffajee AD, Devlin PM, Norris CM, Posner MR, Goodson JM. The salivary microbiota as a diagnostic indicator of oral cancer: a descriptive, non-randomized study of cancer-free and oral squamous cell carcinoma subjects. J Transl Med 2005;3:27.

[15] Byakodi R, Krishnappa R, Keluskar V, Bagewadi A, Shetti A. The microbial flora associated with oral carcinomas. Quintessence Int 2011;42(9):e118–23.

[16] Park SY, Hwang BO, Lim M, Ok SH, Lee SK, Chun KS, et al. Oral-gut microbiome axis in gastrointestinal disease and cancer. Cancers (Basel) 2021;13(9).

[17] Cramer JD, Burtness B, Ferris RL. Immunotherapy for head and neck cancer: recent advances and future directions. Oral Oncol 2019;99:104460.

[18] Shahoumi LA, Yeudall WA. Targeted therapies for non-HPV-related head and neck cancer: challenges and opportunities in the context of predictive, preventive, and personalized medicine. EPMA J 2019;10(3):291–305.

[19] Agulnik M. New approaches to EGFR inhibition for locally advanced or metastatic squamous cell carcinoma of the head and neck (SCCHN). Med Oncol 2012;29(4):2481–91.

[20] Cohen RB. Current challenges and clinical investigations of epidermal growth factor receptor (EGFR)- and ErbB family-targeted agents in the treatment of head and neck squamous cell carcinoma (HNSCC). Cancer Treat Rev 2014;40(4):567–77.

[21] Ishida K, Tomita H, Nakashima T, Hirata A, Tanaka T, Shibata T, et al. Current mouse models of oral squamous cell carcinoma: genetic and chemically induced models. Oral Oncol 2017;73:16–20.

[22] Nakashima T, Tomita H, Hirata A, Ishida K, Hisamatsu K, Hatano Y, et al. Promotion of cell proliferation by the proto-oncogene DEK enhances oral squamous cell carcinogenesis through field cancerization. Cancer Med 2017;6(10):2424–39.

[23] Ishida K, Tomita H, Kanayama T, Noguchi K, Niwa A, Kawaguchi M, et al. Specific deletion of p16(INK4a) with retention of p19(ARF) enhances the development of invasive oral squamous cell carcinoma. Am J Pathol 2020;190(6):1332–42.

Index

Note: Page numbers followed by *f* indicate figures and *t* indicate tables.

A

Aging, 109–110, 109*f*
Alcohol and oral cancer, 1–2, 7
Allopurinol, 187
Annexin A8-like2 (ANXA8), 158
Aspiration pneumonia, 106–107
Autophagy
  aging, 109–110, 109*f*
  autophagosome formation, 102–104, 103*f*
  cancer
    cancer stem cells (CSCs), 114
    chloroquine (CQ), 117–118
    hydroxychloroquine (HCQ), 117–118
    role, 113, 113*f*
    SBI-0206965 (SBI), 117
    therapeutic targets at early stages, 117
    therapeutic targets at late stages, 117–118
    therapeutic tool, 117
    tumor cell dissemination and metastasis, 116–117
    tumor metabolism, 113–114
    tumor-promoting function, 116
    tumor-suppressing function, 115–116
  cellular differentiation, 104–105
  cellular senescence
    ATG7 and ATG5, 111
    biomarkers, 110–111
    description, 110
    external stresses, 111
    p38 mitogen-activated protein kinase α
      (MAPK α), 112, 112*f*
    Rap/3MA treatment, 111
    types, 110
    in vitro and in vivo study, 110–111
  concept, 101–102
  lysosomal system, 101–102
  oral infection
    aspiration pneumonia, 106–107
    oral cavity, 106–107
    xenophagy, 107–109, 108*f*
  oral mucosal homeostasis
    maintenance, 102

    oral epithelium, 101–102
    oral mucosal architecture, 102
    osteogenesis regulation, 105–106
    signaling pathways, 104
    ubiquitin proteasome system (UPS), 101–102

B

Benzydamine oral rinse, 189–190
Betel quid and oral cancer, 7
Bisulfite sequencing method, 85
Brachytherapy, 166, 170–171
Buccal cancer, 64–65, 64–65*f*

C

Cancer stem cells (CSCs), 114
Candida infection, 9
Carcinoma in situ (CIS), 32–33, 33–34*f*
Cellular homeostasis, 101–102
Cellular senescence
  ATG7 and ATG5, 111
  biomarkers, 110–111
  description, 110
  external stresses, 111
  p38 mitogen-activated protein kinase α
    (MAPK α), 112, 112*f*
  Rap/3MA treatment, 111
  types, 110
  in vitro and in vivo study, 110–111
Cervical lymph node metastases, 151–152
CK19 expression, 155–157
Conformal therapy, 172
Cryotherapy, 186
Cytology. *See* Oral mucosal exfoliative
    cytology (OMEC)

D

Desmoglein3 (DSG3), 157
Diagnosis
  oral cancer
    carcinoma in situ (CIS), 32–33, 33–34*f*
    clinical findings, 29

205

206 Index

Diagnosis *(Continued)*
  epidemiology, 29
  histological types, 30
squamous cell carcinoma (SCC) *(see
    Squamous cell carcinoma (SCC))*
  oral epithelial dysplasia (OED)
    classification, 25–26, 26*f*
    definition, 25–26
    diagnostic criteria, 25*t*
    immunohistochemistry (IHC), 26–29,
      27–28*f*
  oral potentially malignant disorders (OPMDs)
    classification, 20, 20*t*
    definition, 20
    discoid lupus erythematosus, 24
    erythroplakia, 21, 22*f*
    factors, 20
    leukoplakia, 20–21, 21*f*
    oral candidiasis, 22–24, 24*f*
    oral lichen planus, 21–22, 23*f*
    oral submucous fibrosis, 22, 23*f*
    oral syphilis, 24
    proliferative verrucous leukoplakia
      (PVL), 25
Discoid lupus erythematosus, 24
Distant metastases, 76
DNA methyltransferases, 82, 82*f*
Dye method, 152–153

E

Epidemiology of oral cancer
  age and sex, 3
  chronic irritation, 1
  genetic mutation and signal transduction
    alteration events, 9–10, 10*t*
    epidermal growth factor receptor (EGFR)
      pathway, 11
    hallmarks of cancer, 9–10, 9*f*
    Hippo pathway, 13
    Jak and STAT pathway, 12
    mTOR pathway, 11
    NOTCH pathway, 12–13
    nuclear factor-kappa B (NF-κB) pathway, 12
    PD-1/PD-L1 pathway, 13
    PI3K and AKT pathway, 11
    RAS and MAPK pathway, 11
    WNT pathway, 12
  histological type, 2–3, 4*t*
  incidence and mortality, 4–5, 5–6*f*
  inflammation and cancer
    cancer microenvironment, 14–15
    causes of inflammation, 14

    chronic inflammation, 14
    immune checkpoint mechanism, 15
    infectious diseases, 14
    tumor immunity, 15
  oral anatomy and favorite site, 2, 3*t*
  risk factors
    alcohol and oral cancer, 7
    betel quid and oral cancer, 7
    Candida infection, 9
    Epstein–Barr virus (EBV), 8
    human papillomavirus (HPV), 7–8
    oral microbiome, 8
    tobacco and oral cancer, 7
Epidermal growth factor receptor (EGFR)
    pathway, 11
Epidermotropism, 139
Epigenetics
  abnormalities, 79–80
  activation of oncogenes, 83–84
  assay method, 91–92
  blood and body fluids, 91–92
  cancer treatment, 90–91
  causes of methylation abnormalities
    endogenous factors, 84, 85*f*
    exogenous factors, 84, 85*f*
  concept, 79
  CpG island standards, 83
  DNA methylation
    in cancer, 83
    CpG islands, 80
    cytosine bases, 81–82, 81*f*
    frequency, 81
    hypomethylation and hypermethylation, 81
  DNA methyltransferases, 82, 82*f*
  epigenetic abnormalities, 79–80
  future research, 98
  hypermethylation detection, 91–92
  mechanisms, 79–80
  methylation detection methods, 86*f*
    comprehensive genome analysis, 87
    individual gene regions analysis, 85
  methylation inhibitors, 90–91
  methylation therapies, 92–93
  myelodysplastic syndrome (MDS), 93
  nonoral cancer
    breast cancer, 89
    colorectal cancer, 90
    ovarian cancer, 89
    ulcerative colitis, 89–90
  oral cancer
    CHFR gene expression, 87–88
    collagen gel invasion model, 97*f*
    demethylation agent DAC, 93–94

Index **207**

DNA methylation abnormalities, 88
dose-dependent and time course, 96*f*
epigallocatechin gallate (EGCG), 94,
95–97*f*, 98*t*
EZH2 inhibitors, 97
HDAC inhibitors, 94–97
invasive foci and invasion depth, 96*f*, 98*t*
M stage checkpoint gene, 87
p16 and MGMT genes, 88, 88–89*t*
RECK gene, 94, 95–96*f*
pharmaceuticals, 92–93
reversibility of methylation, 82
silencing of tumor suppressor genes, 84
studies, 80
Epstein–Barr virus (EBV), 8
Erythroplakia, 21, 22*f*
External beam radiotherapy (EBRT), 166,
171–172, 172*f*

F

Floor of the mouth cancer, 65–66, 66*f*

G

Gingival cancer, 64–65, 64–65*f*
GLABOCAN, 1–2
Glutamine, 188–189
Graft-*versus*-host disease (GVHD)
acute
oral findings, 130–131
symptoms, 128–129, 128*f*
characterization, 127
chronic
oral findings, 131–133
symptoms, 129, 130*f*
development, 127–128
immunopathophysiology
afferent phase, 135
animal models, 133–134
efferent phase, 135–136
mucocutaneous GVHD (*see*
Mucocutaneous graft-*versus*-host
disease)
representation, 134*f*
three step model, 134–136, 134*f*
tissue damage, 135
oral squamous cell carcinoma (OSCC)
donor T cells, 143
graft-*versus*-tumor (GVT) effects,
142–143
haploidentical-SCT (halo-SCT), 143–144
hematologic malignancies, 143
survival rate, 142

H

Hard palate cancer
adenoid cystic carcinoma, 67–69, 68*f*
squamous cell carcinoma, 67, 67*f*
Head and neck cancers, 151
Hippo pathway, 13
Human papillomavirus (HPV), 7–8
Hypermethylation detection, 91–92

I

Image-guided radiation therapy (IGRT),
173–175
Imaging findings of oral cancer
anatomical sites of cavity, 56
buccal mucosa and gingival cancer, 64–65,
64–65*f*
floor of the mouth cancer, 65–66, 66*f*
hard palate cancer
adenoid cystic carcinoma, 67–69, 68*f*
squamous cell carcinoma, 67, 67*f*
imaging modalities and protocols
^{18}F-fluorodeoxyglucose positron emission
tomography, 60
computed tomography (CT), 58, 59*f*
magnetic resonance (MR) imaging, 59–60
limitation, 55–56
lip lesions, 60–61
retromolar trigone, 69, 70*f*
risk factors, 55
spreading patterns
bone invasion, 71, 72*f*
distant metastases, 76
lymph node metastasis, 73–75, 75*f*
perineural spread, 71–73
staging of oral cancer
M categorization, 58, 58*t*
N categorization, 56, 57*t*
T categorization, 56, 57*t*
tongue cancer
DOI, 61
lymphatic spread, 63
squamous cell carcinoma (SCC), 61, 62*f*
tongue musculature, 61–63, 63*f*
treatment, 55
Indefinite for neoplasia (IFN), 47
Infectious diseases, 1
Intensity-modulated radiation therapy
(IMRT), 172–173, 174*f*

J

Jak and STAT pathway, 12

208 Index

L
Leukoplakia, 20–21, 21*f*
Lip lesions, 60–61
Liquid-based cytology (LBC), 43–44, 52–53, 53*t*. *See also* Oral mucosal exfoliative cytology (OMEC)
Lymph node metastasis, 73–75, 75*f*

M
M categorization, 58, 58*t*
mTOR pathway, 11
Mucocutaneous graft-*versus*-host disease
allostimulation, 137–138
apoptosis, 141
cytokines, 137–138
effector T cells, 140
histopathology, 136
homing, 138
ICAM-1 expression, 139, 141–142
lymphocytes in epidermis, 136
lymphoid epidermotropism, 136–137
macrophages, 140, 140*f*
MHC class II expression, 141–142, 141*f*
satellitosis, 136–137
targeting phase, 139
Myelodysplastic syndrome (MDS), 93

N
N categorization, 56, 57*t*
Near-infrared fluorescence imaging, 153
Negative for intraepithelial lesion or malignancy (NILM), 46, 47*f*
NOTCH pathway, 12–13
Nuclear factor-kappa B (NF-κB) pathway, 12

O
One-step nucleic acid amplification (OSNA) assay
annexin A8-like2 (ANXA8), 158
candidate markers, 158
CK19 expression, 155–157
desmoglein3 (DSG3), 157
vs. histopathology, 155
intraoperative genetic diagnostic test, 154
oral cancer, 155–156, 156*t*
p63 biomarker, 157
procedure, 154
sensitivity and specificity, 155
squamous epithelial cancers, 154

Oral cancer
carcinoma in situ (CIS), 32–33, 33–34*f*
clinical findings, 29
epidemiology, 29
epigenetics
CHFR gene expression, 87–88
collagen gel invasion model, 97*f*
demethylation agent DAC, 93–94
DNA methylation abnormalities, 88
dose-dependent and time course, 96*f*
epigallocatechin gallate (EGCG), 94, 95–97*f*, 98*t*
EZH2 inhibitors, 97
HDAC inhibitors, 94–97
invasive foci and invasion depth, 96*f*, 98*t*
M stage checkpoint gene, 87
p16 and MGMT genes, 88, 88–89*t*
RECK gene, 94, 95–96*f*
histological types, 30
imaging findings (*see* Imaging findings of oral cancer)
Oral candidiasis, 22–24, 24*f*
Oral care, 185–186
Oral cavity cancer, 165. *See also* Radiotherapy
Oral epithelial dysplasia (OED)
classification, 25–26, 26*f*
definition, 25–26
diagnostic criteria, 25*t*
immunohistochemistry (IHC), 26–29, 27–28*f*
Oral high-grade squamous intraepithelial lesion of low-grade epithelial dysplasia (OHSIL), 47, 48*f*
Oral lichen planus (OLP), 21–22, 23*f*, 199–200
Oral low-grade squamous intraepithelial lesion of low-grade epithelial dysplasia (OLSIL), 46–47, 48*f*
Oral microbiome, 8
Oral mucosal exfoliative cytology (OMEC)
as adjunctive test, 43
cell collection procedure, 44
cytological classification
basic procedures, 51, 51*t*
cytodiagnostic flowchart and treatment plan, 46*f*
diagnostic criteria, 46*t*
differential cytodiagnosis, 52
histological grading system, 49–50, 50*t*
indefinite for neoplasia (IFN), 47
negative for intraepithelial lesion or malignancy (NILM), 46, 47*f*

Index **209**

oral epithelial dysplasia (OED), 49–50, 50*t*
oral high-grade squamous intraepithelial
lesion of low-grade epithelial dysplasia
(OHSIL), 47, 48*f*
oral low-grade squamous intraepithelial
lesion of low-grade epithelial dysplasia
(OLSIL), 46–47, 48*f*
squamous cell carcinoma (SCC), 47, 49*f*
usefulness, limitations, and future
research, 52–53, 53*t*
diagnostic criteria, 45
oral mucosa structure and function, 43–44
specimen evaluation, 45
Oral mucositis (OM)
clinical presentation, 183
developmental process, 185, 185*f*
dose-limiting toxicity, 183
frequency, 184
management
allopurinol, 187
benzydamine, 189–190
cryotherapy, 186
glutamine, 188–189
oral care, 185–186
palifermin, 190–191
sucralfate, 187–188
vitamin E, 186–187
zinc sulfate, 189
pathogenesis, 184–185, 185*f*
polaprezinc effect, 191–192
severe oral mucositis, 183
Oral potentially malignant disorders (OPMDs),
199–200
classification, 20, 20*t*
definition, 20
discoid lupus erythematosus, 24
erythroplakia, 21, 22*f*
factors, 20
leukoplakia, 20–21, 21*f*
oral candidiasis, 22–24, 24*f*
oral lichen planus, 21–22, 23*f*
oral submucous fibrosis, 22, 23*f*
oral syphilis, 24
proliferative verrucous leukoplakia (PVL), 25
Oral squamous cell carcinoma (OSCC), 43
adjuvant radiotherapy with chemotherapy,
201–202
basic and clinical research, 203, 203*f*
biomarkers, 200–201
cancer stem cell markers, 200–201
immune checkpoint inhibitors, 202
incidence rate, 199

liquid biopsies, 201
non-OSCC neoplasms, 200
oral hygiene, 201
oral microbiomes, 201
oral potentially malignant disorders
(OPMDs), 199–200
preclinical data, 201
risk factors, 199
surgery, 201–202
treatment planning, 202
treatment responses, 202–203
Oral submucous fibrosis, 22, 23*f*
Oral syphilis, 24
Osteogenesis, 105–106

P

p63 biomarker, 157
Palatal tumor, 67–69, 67–68*f*
Palifermin, 190–191
Particle beam therapy (PBT), 175–176,
175–176*f*
PD-1/PD-L1 pathway, 13
PI3K and AKT pathway, 11
Polaprezinc effect, 191–192
Postoperative radiotherapy (PORT)
multiple lymph node metastases, 167
postoperative chemoradiotherapy (POCRT),
167
recurrence risk factors, 166–167
treatment outcomes, 167, 168*t*
Premature senescence, 110
Proliferative verrucous leukoplakia (PVL), 25,
199–200
Prophylactic cervical lymphadenectomy,
151–152

R

Radioisotope (RI) method, 152–153
Radiotherapy
brachytherapy, 166, 170–171
definitive radiotherapy, 166, 166*t*
early-stage oral cavity cancer, 166, 166*t*
external beam radiotherapy (EBRT), 166,
171–172, 172*f*
functional organ preservation approaches, 169
general principle, 166*t*
image-guided radiation therapy (IGRT),
173–175
intensity-modulated radiation therapy (IMRT),
172–173, 174*f*
oral cavity cancer classification, 165

210 Index

Radiotherapy *(Continued)*
  palliative radiotherapy, 169–170
  particle beam therapy (PBT), 175–176,
    175–176*f*
  postoperative radiotherapy (PORT)
    multiple lymph node metastases, 167
    postoperative chemoradiotherapy
      (POCRT), 167
    recurrence risk factors, 166–167
    treatment outcomes, 167, 168*t*
  primary radiotherapy/chemoradiotherapy, 169
  surgical resection, 165
  toxicity management, 170
RAS and MAPK pathway, 11
Replicative senescence, 110
Retromolar trigone, 69, 70*f*
Risk factors, 55

S

Sentinel lymph node biopsy (SLNB)
  application, 152
  cervical lymph node metastases, 151
  evidence-based medicine (EBM), 158–159
  vs. imaging systems, 151–152
  metastasis diagnosis
    diagnostic accuracy, 154
    one-step nucleic acid amplification
      (OSNA) assay (*see* One-step nucleic
      acid amplification (OSNA) assay)
  vs. prophylactic cervical lymphadenectomy,
    151–152
  sentinel lymph node identification methods
    ^{99m}Tc-tilmanocept, 153–154
    dye method, 152–153
    near-infrared fluorescence imaging, 153
    radioisotope (RI) method, 152–153
    SPECT/CT, 154
  wait-and-see approach, 152
Squamous cell carcinoma (SCC), 47, 49*f*
  acantholytic squamous cell carcinoma, 39, 39*f*
  adenosquamous carcinoma, 36

basaloid squamous cell carcinoma, 35, 35*f*
carcinoma cuniculatum, 36–37
grade classification, 31–32
histology, 30–31, 30*f*
immunohistochemistry (IHC), 32
lymph node metastasis, 30–31, 31*f*
lymphoepithelial carcinoma, 38
papillary squamous cell carcinoma, 38, 38*f*
spindle cell squamous cell carcinoma,
  35–36, 36*f*
verrucous squamous cell carcinoma, 37, 37*f*
Sucralfate, 187–188

T

T categorization, 56, 57*t*
^{99m}Tc-tilmanocept, 153–154
Tobacco and oral cancer, 1–2, 7
Tongue cancer
  DOI, 61
  lymphatic spread, 63
  squamous cell carcinoma (SCC), 61, 62*f*
  tongue musculature, 61–63, 63*f*

V

Vitamin E, 186–187

W

Warburg hypothesis, 113–114
WNT pathway, 12

X

Xenophagy
  description, 107
  periodontal infection, 108–109, 108*f*

Z

Zinc sulfate, 189

Printed in the United States
by Baker & Taylor Publisher Services